UNMENTIONABLE SUBJECTS

A PLAIN MAN'S GUIDE TO THE STRANGE WORLD OF REALITY

"Beauty is truth, and truth beauty."

Keats

"It is almost as if the utter insignificance of the Earth and the thin film of its atmosphere has totally paralysed the imagination, as if it were too dreadful to contemplate and therefore best ignored."

Francis Crick, Nobel prizewinner

"Let us hope it is not true, but if it is true then let us hope it will not become generally known."

attributed to wife of *Bishop Wilberforce*

"Humankind cannot bear very much reality."

T. S. Eliot

"Human beings engage in wilful deception on a massive scale. The ability to tell lies is perhaps our most striking characteristic."

Professor Sir Edmund Leach, leading anthropologist

"Primitive people, to whom many aspects of the physical world are terrifying (because not understood), fill their environment with imaginary gods, ghosts, demons and monsters."

Isaac Asimov

"A purge of obscure Chinese gods and goddesses has been launched in Taiwan by the Chinese Nationalist authorities. Forty-five gods and goddesses are honoured in most Taipei temples. Mr. Huang Yu-yuan, Director of Civil Affairs in Taipei said, "In this way men create gods only to fool ignorant people." "

news item in *Times* 17.7.'82

"In the long run it is far more dangerous to adhere to illusion than to face what the actual fact is. And secondly, what is the point of life if you live in an invented world, if there is no relationship either to the world or people or anything. It isn't a relationship if you are related to something which isn't there or which is just there to make yourself feel comfortable. I think we all have this tendency and it is the problem of mankind . . . this is the thing which has always made mankind's life miserable and which is threatening our survival. We cannot face the implications of what we are doing and it will lead us over the edge into the abyss."

Professor David Bohm, world renowned physicist

Published in 1986 by
VANGUARD BOOKS HEREFORD

ISBN 0 9511115 0 7

Printed in England by
A Wheaton and Co. Ltd., Exeter

Contents

PHOTOGRAPHS AND ACKNOWLEDGEMENTS

CHAPTER I (between pages 30–31)

ANDROMEDA GALAXY	(*Science Photo Library*)
EDGE-ON SPIRAL GALAXY	(*Science Photo Library*)
ORION NEBULA	(*Science Photo Library*)
CRAB NEBULA	(*Science Photo Library*)
GOLD ATOMS/COSMIC RAY	(*Science Photo Library*)
MATTER ANTIMATTER CREATION	(*Science Photo Library*)
ALBERT EINSTEIN	(*Hulton Picture Library*)
APOLLO MOON ROCKET	(*Hulton Picture Library*)

CHAPTER III (between pages 62–63)

PHAGE VIRUS/BETA PICTORIS	(*Science Photo Library*)
PHAGE VIRUS × 250,000	(*Science Photo Library*)
A. AFRICANUS SKULL	(*Nat. Museums of Kenya*)
HOMO HABILIS SKULL 1470	(*Nat. Museums of Kenya*)
ARCHAEOPTERYX FOSSIL	(*Ardea Photographics*)

CHAPTER VI (between pages 142–143)

CONQUISTADORE ATROCITIES	(*Hulton Picture Library*)
JONESTOWN MASS SUICIDE	(*Hulton Picture Library*)
ADOLF HITLER	(*Hulton Picture Library*)
NAZI BOOK-BURNING	(*Weiner Library Archive*)
CAMBODIAN KILLING FIELDS	(*Hulton Picture Library*)
DRESDEN FIRESTORM	(*Imperial War Museum*)
DRESDEN AFTER THE AIR-RAID	(*Hulton Picture Library*)
CONCENTRATION CAMP SURVIVORS	(*Museum of Boer Republics, Bloemfontein*)

Foreword

A BOOK on human anatomy which left out the unmentionable parts would by definition not be the best book possible, and in this respect the anatomy of reality is the same: we must look at all the parts if we wish to understand the whole. The claim that a wide diversity of evidence is inadmissible and reality, if it is to be explained at all, must be explained in terms of particle physics, Zen buddhism or some similarly limited and homogeneous philosophy may be comforting but has little to do with truth. The argument against a diversity of evidence would certainly not be accepted in any court of law unless the criminals were running the court. Like the criminals, human nature finds the abundance of evidence embarrassing—as well it might.

This book was written for all who have ever wondered "Why?" or "How?" and, in a slightly different sense, for all those who believe reality can be anything they want it to be. You will note, perhaps disapprovingly, that there are times when I find it hard to suppress a sense of anger, not only at the Pizarros, the Idi Amins and Pol Pots of this world, their imitators and kindred spirits both large and small, but at the lemming hordes who by their blind faith and mindless conformity make such iniquity possible. If I have failed to rouse this anger in my reader I have not done my job properly, or else he is more cold a fish than he has any right to be. Worse still, he may himself be suffering from that over-abundance of human nature which has been the hall-mark of tyrants and criminals through the centuries; those who feel no anger at the perpetrators of injustice and atrocity are almost certainly fellow travellers and kindred spirits. Loathsome creatures of whatever kind do not normally regard one another as loathsome.

We are about to embark on an exploration into the strange world of reality, so different from our everyday experience, yet hidden from us more by our own dishonesty than by any lack of access to the evidence. Human nature is the glass through which we are forced to view the world around us, and to accept what we see without taking into account the imperfections and distortions of the medium through which we see it would be a major folly.

It does no good to look the other way and pretend it does not matter—it does matter, and if we cannot bear to look at ourselves except through rose-tinted spectacles, what kind of people are we? The repulsive side of human nature, present in every one of us, will not go away if we ignore it. On the contrary, it will flourish as never before, distorting our perception and creating an enormous discrepancy between what we see and what is actually there. To thus allow weakness and self-indulgence to interfere with truth has obvious moral implications, especially as failure to apply the necessary corrective measures will lead to a series of avoidable human disasters on a scale which will make the hazards of radiation and industrial pollution seem trivial by comparison. Not perhaps, or maybe, but inevitably. As the record shows, human nature is the most dangerous pollutant of them all.

Reality is certainly very interesting, often exciting, but definitely not comforting or cosy, and the more we try to pretend it is the less cosy it becomes. A significant part of reality's harshness is caused by mankind's attempts to evade it, and our make-believe cosiness is bought at a fearful price. Truth sometimes hurts, of course, but will only be unbearable agony to people who hate it. And while truth is only uncomfortable in the emotional sense, make-believe often brings physical discomfort as well. Your discomfort, whether physical or emotional, is often the result of someone else's make-believe. It is an intriguing paradox that the more readily we admit that reality is harsh and contrary to our emotional requirements, the less harsh it will be. Our emotional requirements are in any case partly artificial, created by incompetent programmers rather than our genetic inheritance.

To know reality is to know ourselves. No wonder some people would rather not look! There is a way out, but it is painful—though not nearly so painful as the alternative—and the longer we leave it the harder it will be. If we leave it too long it will be impossible. The saddest words in the English language, it is said, are the words 'too late'. Ignorance of reality and hatred of truth always was a dangerous business, and countless millions have paid for it with their lives—often, unfortunately, not the people who should have paid. But times have changed. The world has become far too dangerous to allow reality to remain taboo any longer. The kidding has to stop, and much of the killing will stop with it. It is now no longer reality which is too awful to contemplate, but the alternative.

There can be few things more dehumanising than belief in magic, which by its very nature will tend to stifle any spirit of enquiry. Once a mystery or phenomenon has been shrugged off as due to magic there is by definition no incentive to look further for the true explanation, especially as this might severely tax grey matter unused to such exercises, or shatter cherished illusions. To freely adapt a much quoted line from the Hindu Bhagavad Gita, ". . . I am become truth, the shatterer of worlds!"

Reality is not in the mind but external, and exists quite independently of human thought or even human existence. It will still be here, remorseless and uncaring as always, when we have gone to join the dinosaurs. Contrary to a widely held belief there is only one ultimate reality and, however harsh it might seem, there are no variations or alternatives. The variations which we think we see arise not in reality itself but in our different perceptions and interpretations of it. The brain does not exist to create reality but is there to reach out and probe it with the sensory organs, analysing the data in order to build models, make predictions, and co-ordinate our responses so that we may survive to propagate our kind.

Ultimate reality, like a temperature of absolute zero or the speed of light, can be approached ever more closely but never actually attained, and each degree of closeness will resist attack more stubbornly than the one before. But regardless of those who would have you believe otherwise, a determined attack will succeed far more than a half-hearted attack or no attack at all. Reality is somewhere out there, challenging us to explore, daring us to look. The greater our understanding the longer our departure into oblivion can be delayed.

In any good translation, censorship must be avoided. Although there is room for some small variation of emphasis and opinion, honesty demands that our interpretation of reality, as with any other form of interpretation, should be as close to the original as possible. It follows that no meaningful interpretation will be possible unless we examine the original text, which nature reveals to us in a multiplicity of natural phenomena. Those who claim to offer a translation without having examined these phenomcna in detail (and sometimes with no knowledge of them at all) are deceiving both themselves and you. Anyone who finds nature's original text uninteresting or even repulsive will not make a good translator and should be honest enough to admit it.

Finally, to those who may feel I have been too outspoken, I say this: if you do not get plain speaking from me, who will you get it from? Truth is not well served either by censorship or by beating about the bush. Time is running out. "Humankind cannot bear very much reality" wrote T. S. Eliot. Unless we change our ways very quickly, we may find that reality cannot bear very much mankind. In so far as only part of the available evidence has been quoted and much useful ammunition left unfired, the following chapters are an understatement.

I

The Eternal Universe

We live on a tiny speck of matter whirling through space and time: the Earth as it orbits the mighty Milky Way galaxy once every 250 million years. The galaxy itself, though 100,000 light years in diameter, is but one of billions, and no more than a sand grain in the Sahara of the universe. The three score years and ten of the average human lifespan is but the bat of an eyelid. It follows that in order to be accommodated within a timespan which can be even remotely related to human lifetimes and encompassed within the dimensions of that tiny speck of cosmic dust, all things of our Earthly experience will be finite.

The idea of infinity in either time or space has therefore always been difficult for man to come to terms with except as a mathematical abstraction, or in cyclic or circular forms in which beginnings and ends remain unclear. This is because infinity is outside our everyday experience. Even the idea of a million, a small enough figure by astronomical or geological standards, is difficult for most of us to grasp, despite the way it trips so lightly and easily off the tongue. To put it in proper perspective, it will be another 750 years before a million days have elapsed since the birth of Christ. About 800 more years of sunrises must come and go before the sun will have risen a million times since the Roman conquest of Britain. The idea of a billion (a thousand million) is even harder to comprehend.

Not surprisingly, it was reasoned in ancient times that the universe as a whole would conform to the experience of our daily existence, and would likewise have a beginning and an end. The logic of this is decidedly shaky, and the naively anthropomorphic idea that the universe and everything therein was magically created by some omnipotent supernatural entity is reminiscent of the ancient belief that the Earth is supported on the back of a giant turtle swimming in a bowl, which is in turn supported on the back of an elephant with golden tusks, which in turn stands on . . . what? The original problem has been brushed under the carpet only to be replaced by another. Far from solving anything, such childish solutions merely add unnecessary layers of complexity for which there is no good evidence.

Just as one might reasonably ask what supported the elephant with the golden tusks, the enquiring mind must now ask: from whence came this omnipotent deity or deities? There are only two possible answers. Either they were created by yet another all-powerful being (an absurd process which could go on ad infinitum) or else they are themselves eternal and have always existed. So you see, the idea of an eternal entity is inescapable whatever view of the universe you take. Just what the various deities did to occupy their time before creating the universe, like what supported the elephant with the golden tusks, is not explained by believers. It is far more rational and in greater accord with scientific principle to say that this eternal entity was the universe itself rather than any system of creators, thus removing at a stroke the unnecessary complexities which destroy the elegance of more traditional theories. Most people find a never ending universe with infinite existence into the future relatively easy to accept. The proposition of a universe with no beginning, extending infinitely into the past, is no less logical.

At the dawn of time the black stillness of the void was shattered by an explosion of unimaginable violence as radiant energy from another dimension in space and time blasted into the here and now. The 'Big Bang', a colossal white hole from which a fiery new universe sprang phoenix-like from the ashes of the old, had occurred. It was not until some fifteen billion years later that man was able to piece together the evidence which pointed back to this dramatic event.

Sir Isaac Newton, perhaps the greatest scientist of all time, first showed white light to be a rainbow mixture of colours when he diffracted sunlight through a prism in the seventeenth century. Because the short wavelength of violet is diffracted or bent more easily than the long wavelength of red, the light is broken into its component parts. The resulting bands of colour, similar to a rainbow and ranging from red through orange, yellow, green and blue to violet, are known as a spectrum.

Later investigators showed that any element in gaseous or vapour form heated to brilliant incandescence gives off light of certain well defined wavelengths which are characteristic of that element, and show up as bright lines at precise points in the spectrum. When solids or very dense gases are heated, these bright emission lines merge to form the continuous spectrum of white light.

However, the same atoms which emit this bright line emission 'signature' when incandescent will also absorb light at these wavelengths when too cool to emit or when light from a brighter source passes through them. Therefore the spectrum of sunlight is not perfectly continuous but crossed with a few dark lines at certain places where the gases of the sun's corona or outer atmosphere have absorbed their characteristic 'signature' wavelengths and filtered them out of the spectrum. These lines, by their position in the spectrum, indicate the identity of the absorbing gases and are called absorption lines.

Not only the sun but all other stars have these dark lines in their spectra, and their position and strength indicate which gases are present in the star's atmosphere. Much other useful information can also be deduced from these absorption lines. If the lines characteristic of, say, hydrogen are shifted towards the red end of the spectrum, this indicates that the light source is moving away from the observer, and is called a red-shift. When displacement is towards the blue, this indicates motion towards an observer and is known as a blue-shift.

This displacement is due to the Doppler effect, which also causes the whistle of a train to sound high pitched when approaching, dropping to a lower pitch as it passes an observer and moves away. When the effect applies to light waves instead of sound waves, a blue-shift is analogous to the high pitch of the whistle as the train approaches and a red-shift to the lower note as the train moves away.

In 1920, V. M. Slipher, working at the Flagstaff Observatory in Arizona, discovered that all galaxies (gigantic discus-shaped star systems) except the local group had red-shifted absorption lines in their spectra. Later in the nineteen-twenties the renowned astronomer Edwin Hubble developed techniques for estimating the distances of galaxies, and discovered a curious relationship between these distances and their red-shifts. It turned out that the further away a galaxy was the greater would be its red-shift, and the variation was proportional to the distance of the galaxy from our own star system, the Milky Way, which is itself a galaxy.

As red-shift is produced by recession, it seemed that all the galaxies were receding from us and from one another, some (the more remote ones) at tremendous velocities of thousands of miles per second. The universe seemed to be flying apart!

One of the first to offer an explanation for this puzzling state of affairs was a Belgian priest, the Abbé Lemaitre. He reasoned that if you went backward in time, played the film of the universe in reverse as it were, the galaxies would all converge at a central point which would be their point of origin. The Abbé's theory was that many millions of years ago all the matter and energy in the universe was concentrated at this central point in a gigantic super-dense sphere which he called the primeval atom. After an eternity of stable existence, this primeval atom suddenly exploded with incredible violence, throwing out the galaxies in all directions like pieces of shrapnel. They have been flying apart ever since, and this explains the expanding universe which we see around us today.

This theory has since been refined and elaborated by other astronomers and physicists, principally the Russian/American physicist George Gamow, but before explaining these more modern ideas it is necessary to give a few details of the structure of the atom. The study of the infinitely large is linked in a fascinating way with the study of the infinitely small.

A good description of the atom can be given using comparatively few fundamental particles, despite the fact that many more are known to physicists. Most of the latest sub-atomic particles to be discovered are very short lived (billionths or trillionths of a second) and to some extent may be considered as laboratory artifacts created by the energy of collision.

The simplest of all atoms is that of the lightest element, hydrogen. The hydrogen atom consists of a single positively charged particle, the proton, forming a central nucleus around which orbits a very much smaller and negatively charged particle, the electron. This arrangement produces something like an incredibly tiny solar system, far too small to be seen with even the most powerful electron microscope, in which the nucleus is like a central sun circled by the planetary electron.

Strangely enough, although the electron is only about a two-thousandth of the mass of the central proton, their electric charges though opposite are equal in strength, thus making the atom as a whole electrically neutral. It is the attraction of these unlike charges, like the opposite poles of two magnets, which attracts the electron to the nucleus. Gravity is so weak at a

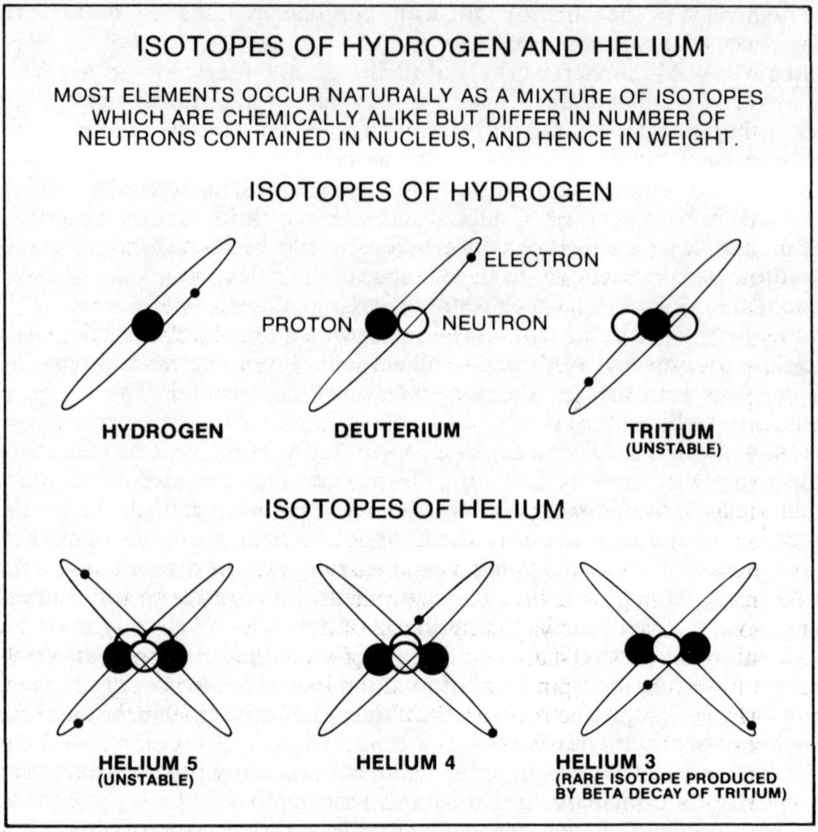

sub-atomic level that it does not come into the picture.

Under certain circumstances the positively charged proton can absorb the negatively charged electron (plus a certain amount of energy) to beome a new particle, the neutron, which is electrically neutral. The neutron can link up with the proton to form a much heavier nucleus which will, however, have only one positive charge and therefore be able to attract only a single orbiting electron.

This heavier form of the atom is known as an isotope, and is chemically similar to the lighter form. The heavy form of hydrogen described above is the isotope deuterium, which, combined with oxygen, forms the well known 'heavy water' used in nuclear research. About 0·02 percent of ordinary tap water is heavy water, but greatly diluted and difficult to separate. There is also a third form of hydrogen, the isotope tritium, which has two neutrons in its nucleus but still only a single proton, and therefore has only one electron. Tritium is unstable and therefore radioactive, decaying by beta emission with a half life of about $12\frac{1}{2}$ years (to become helium 3).

Those of you who find a discussion of the atom heavy going, please bear with me. It will not take long and will give a better understanding of what comes later. Besides, the world of the atom is probably simpler and more interesting than you think. Once you have grasped the idea of symmetry, a theme which occurs time and time again in nature, what formerly seemed complex and incomprehensible will acquire a new and beautiful simplicity. And any explanation in physics, chemistry, astronomy or biology (or anything else for that matter) not only has to be consistent within its own framework but also compatible with our explanation of the atom. If it is not, there is something wrong somewhere—a useful test for truth.

We have so far dealt with three sub-atomic particles which together give an almost complete picture of the atom good enough for most practical purposes. In all other atoms except the common form of hydrogen the protons are joined in the nucleus by equal or greater numbers of neutrons. The main function of the neutron, which has no charge itself, seems to be in allowing the positively charged protons to co-exist in the nucleus without flying apart. Like charges repel, just as two magnets will repel each other if two north poles or two south poles come together—a fact which you can easily test for yourself.

Protons and electrons are stable particles and able to exist indefinitely without decaying (breaking up), whether inside the atom or as free particles. Strangely enough the neutron is only stable within the nucleus and free neutrons decay in about 12½ minutes, breaking down into a proton and an electron and giving off a tiny, almost undetectable particle called an anti-neutrino which streaks away at the speed of light.

Sub-atomic particles share some or all of a number of properties which include mass, charge, spin and magnetic moment. Mass is very variable, ranging from 1 in the electron to 1836 in the proton and 1839 in the neutron. It will thus be seen that in order to become a neutron the proton must gain mass-energy to the tune of three electron masses, and for a neutron to become a proton the reverse must happen.

Charge, which can be either positive or negative, varies very little. It is either present or absent. Spin may be compared to the rotation of the sun, or the Earth, or a bullet, or a cricket ball, and is also called angular momentum. In the world of the atom it comes only at certain well defined values or multiples thereof, and is subject to relatively little variation. Magnetic moment is a consequence of charge, which causes a charged particle to behave like a tiny dipole magnet. The magnetic properties of the neutron suggest that although nominally chargeless it actually contains two opposite charges which cancel each other out. There are a few other more subtle properties of sub-atomic particles which are a bit too technical to go into here.

An extraordinary property of matter at a fundamental level is that every particle has an equal but opposite anti-particle. Each anti-particle is in certain respects the mirror image of its particle counterpart, i.e. their masses are the same but their spin, electric charge, and some other properties are equal but opposite. Einstein showed in his Special Theory of Relativity as far back as 1905 that matter and energy are two different forms of the same thing, and this explains why, when particle and anti-particle encounter each other, both are annihilated to produce an equivalent amount of energy.

Thus when a negatively charged electron meets a positively charged anti-electron (also known as a positron) there is mutual annihilation which results in the production of two gamma-ray photons. The photon is the particle of light and other radiations of the electromagnetic spectrum, which range from the ultra-energetic 'hard' gamma radiation on the one hand, through x-rays, ultra-violet and visible light, to infra-red, microwaves and radio on the other. The energetic waves at the blue end of the spectrum are of very short wavelength which becomes longer and less energetic towards the red end of the spectrum, with the very long wavelength radio waves being the least energetic of all.

It is of considerable significance that not only can one form of energy be converted into another (kinetic into electromagnetic, for example, or vice versa) but in the language of particle physics the masses of particles are routinely described in units of energy (electron volts). The mass of the electron at rest is about half a million electron volts, the same as the energy of the gamma ray created by its destruction.

You may have noticed and been puzzled by the fact that I speak of light first as a particle and then as a wave phenomenon. This paradox is known as wave/particle duality and is best seen in electromagnetic particles, or quanta, which all travel at the speed of light and have no mass. (Neither do they have antiparticles, or to put it another way, they are their own antiparticles, emitted by matter and anti-matter alike.) But it is also seen in matter particles which do have mass, such as the electron. Generally speaking, the larger and slower a particle is the more apparent are the particle aspects of its nature,

THE ELECTROMAGNETIC SPECTRUM

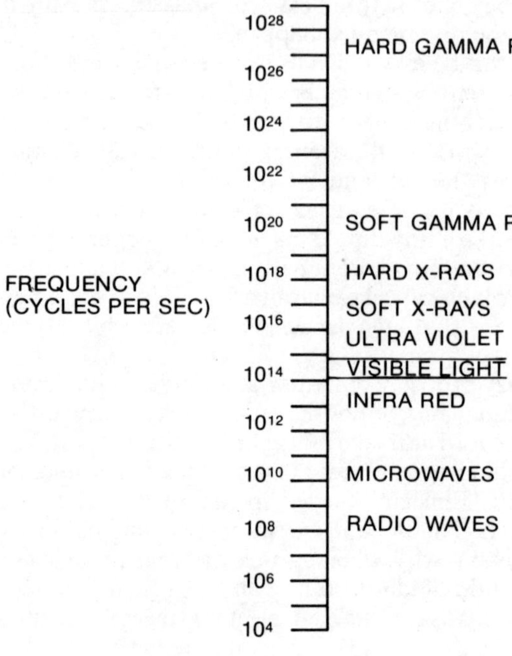

but the smaller and faster it is the more apparent are its wave characteristics.

It is this wave particle duality which makes possible the electron microscope, where beams of electrons take the place of beams of light quanta. Ordinary light microscopes suffer from the limitation that when the objects to be viewed become so small that their size approaches the wavelength of visible light there is a substantial loss of resolution because the light waves are able to bend around them, producing just a blurred and useless image.

The wavelength of an electron (its de Broglie wavelength, named after the man who first explored the wave aspects of particles) is very much smaller than the wavelength of visible light, and because electrons are negatively charged they can be focussed by means of magnets. The object to be studied is bombarded by a beam of electrons accelerated to very high speed. Remember that the smaller and faster the particle, the more apparent are the wave-like aspects of its behaviour.

The electron is very small indeed, only about two-thousandth of the mass of a proton, which is itself extremely minute and cannot be seen in any microscope. In the electron microscope it is also travelling very fast, and this is why a beam of electrons can be used as a beam of light. There are a number of disadvantages, however, one of which is that the object to be viewed is in effect bombarded by a hail of tiny bullets, and this causes damage, especially if the object is alive. In addition, specimens must usually be placed under high vacuum, which is both inconvenient for the operator and injurious to living organisms.

Despite some disadvantages, these are more than offset by the very much higher magnifications obtainable, up to a hundred times higher than those obtainable with ordinary light microscopes. In theory it is even possible to build a proton microscope in which a beam of protons takes the place of a light beam, and this illustrates the very real nature of wave particle duality. It would seem that the apparently solid world we see around us consists of nothing more than ripples in various force fields! It certainly is not solid, as even the heaviest atom is more than 99·999 percent empty space.

It is very difficult to visualise this state of affairs in terms of our everyday experience, as in the macro world apparent to human eyes something is either a particle, like a poppy seed, a salt crystal or a sand grain, or a wave like a sound wave or ripples on a pond. There is no familiar example where it can be both. In the ultra-microscopic world of the atom, however, things are different. Bundles of light waves, also known as photons or light quanta, being extremely small and moving at 186,000 miles per second, usually manifest their wave characteristics more than their particle aspects, but are quite able to do both. They may also be thought of as packets of energy, and in certain situations such as the photo-electric effect, where they knock electrons out of the structure of atoms, they behave like tiny bullets or particles.

They are rather like the man who helps an old lady across the street and then walks off with her purse—one moment showing one side of his character, the next moment another, though he was the same person all the time and did not really change.

It would seem reasonable to suppose that if Einstein was right and matter and energy are one and the same, it should be possible not only to convert matter and anti-matter into energy but also to produce the opposite reaction where energy is converted into matter and anti-matter. This is indeed the case and only one of many such proofs that Einstein was correct.

When a gamma-ray photon (or energy packet) encounters a suitable force field, as in the vicinity of an atomic nucleus, it materialises into two particles—one of matter and the other of anti-matter. Usually these two particles are electrons, or rather an electron and its anti-particle, the positron, and the phenomenon is known as materialisation or pair production. In order to materialise into an electron positron pair a gamma-ray photon must have an energy of at least a million electron volts. On Earth it is not long before the positively charged positron is attracted to a negatively charged electron and the two annihilate each other in a burst of gamma radiation.

Given a gamma-ray photon of sufficient energy (nearly two thousand times stronger than the one which produced electrons) it would be possible to create a proton together with its anti-particle in the laboratory, though in practice anti-protons are created in accelerators with the energy of particle collisions. Allow the positron (or anti-electron) to go into orbit around the anti-proton and you have an atom of anti-hydrogen, the basis of anti-matter so beloved of science fiction writers. But as we have seen, anti-matter, far from being a wild figment of the imagination, is very much a reality. One reason for its great rarity on Earth is that it is not long before anti-electrons and anti-protons are attracted to their oppositely charged counterparts, resulting in mutual annihilation. An anti-neutron, having no charge, might last a little bit longer, but like ordinary neutrons it is unstable outside the nucleus and spontaneously decays in around 12½ minutes. Anti-deuterium and anti-helium 3 have actually been created in the laboratory!

This matter anti-matter annihilation is the most powerful nuclear reaction known, as it converts the particles involved one hundred percent into energy according to Einstein's celebrated equation: energy (in ergs) = mass (in grammes) × speed of light squared ($E = mc^2$ where c equals the speed of light in centimetres per second). This compares with just over half of one percent conversion of mass to energy in the nuclear fusion reaction which powers the hydrogen bomb,† and a very much lower efficiency in 'ordinary' atomic bombs which work by fission (splitting) of heavy and unstable atoms such as uranium 235 or plutonium.

So far we have examined the structure of only the very simplest atom, hydrogen, and its isotopes deuterium and tritium. When atoms are heated they begin to dart wildly about—just as you would if you stepped onto a hot surface in bare feet! The centre of the sun is very hot indeed—several million degrees centigrade—and under these conditions hydrogen atoms dash about at very high speeds. Because of the tremendous pressure they are

† An 'ideal' hydrogen bomb which utilised ordinary hydrogen. Actual hydrogen bombs work by a roundabout route involving lithium, so the yield per pound of fuel is correspondingly reduced.

very closely packed, and consequently there are frequent collisions. When four atoms of hydrogen (protons) collide at sufficiently high speed to overcome the mutual repulsion of their identical positive charges, two of them are converted into neutrons and fuse together with the remaining two protons to form an atom of helium. We here have the transmutation of elements—in essence what the alchemists of old were trying to achieve and so change base metals into gold. (This is theoretically possible in a modern nuclear reactor, but the value of the gold would be less than the cost of production.)

The strange thing is that the mass of the newly created helium atom is less than the combined masses of its four constituent particles by a factor of more than half of one percent. It seems that four times twenty-five equals $99\frac{1}{2}$ when we would expect it to equal 100. So what has happened to the missing mass? It has been converted to energy, and fortunately for us the sun loses mass in this way at the astounding rate of four million tons per second in the form of radiant energy which is dissipated into space. This has been going on for about five billion years and is likely to continue for at least another five billion.

The reaction is usually slightly more complex than that outlined above, moving through stages involving the hydrogen isotopes deuterium and tritium, another route involving helium 3, as well as an even more round about route known as the carbon cycle, but all leading to the same thing in the end. The helium atom is not much more complex than the atom of hydrogen, having a nucleus consisting of two neutrons and two protons. As only the two protons are positively charged, the atom can attract only two negatively charged electrons into orbit around the nucleus.

Electrons in atoms are arranged in distinct 'shells' with no intermediate orbits between them. The innermost shell is the smallest, as might be expected, and therefore holds the fewest electrons—only two. The helium atom has only this one shell (the K shell), which is thus filled to capacity by its two electrons. As atoms combine to form molecules by sharing electrons and electron shells with other atoms, this and the fact that these innermost electrons are the most stable and tightly held explains the inertness of helium, its reluctance to react and form compounds with any other substance. The chemical properties of an element are determined mainly by the outer electron shell.

Remember that the solidity which we associate with objects such as pebbles and billiard balls is an artifact of the macro world created by the combined effects of trillions of sub-atomic particles, and is not duplicated in the structure of the particles themselves. At the atomic or particle level nothing actually touches (there is empty space between your hand and this book, for example) but the mutual repulsion of like charges and the powerful electromagnetic force create an illusion of contact. Atoms, atomic nuclei and molecules should be thought of not so much as rigid and motionless billiard balls but more like drops of water on a red hot iron, dancing and quivering with energy.

In the same way that collisions between hydrogen nuclei result in the formation of helium, so collisions between helium nuclei (which are also called alpha particles) result in the formation of still heavier atoms in a stage

TYPES OF MOLECULAR BONDING

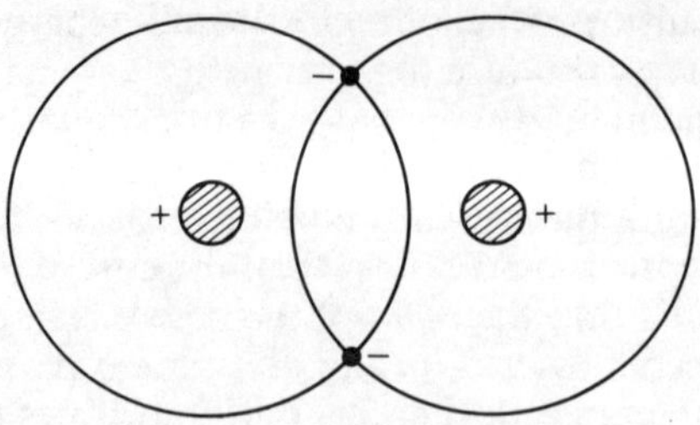

COVALENT BONDING IN HYDROGEN (H_2) MOLECULE

The shared electrons give the effect of full electron shells to both atoms. Because of the attraction of unlike charges the pull of the two hydrogen nuclei on the electrons in between them is sufficient to bond the atoms together.

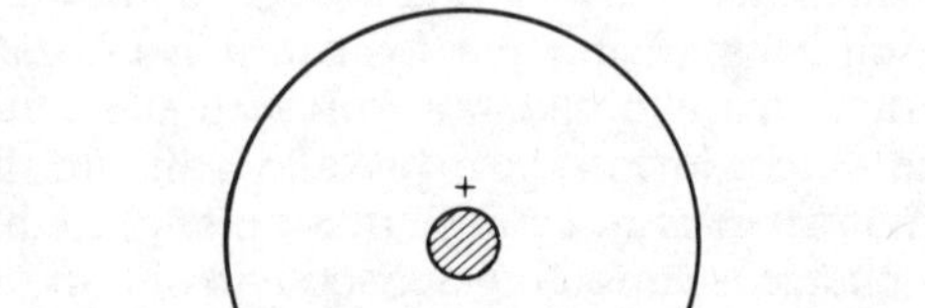

IONIC BONDING IN LITHIUM HYDRIDE

Hydrogen again fills its incomplete K shell, this time by 'stealing' the loosely held outer electron of lithium. Negatively charged hydrogen bonds to positive lithium. Broken line indicates that the outer shell of lithium is now empty, having donated its electron to hydrogen atom. (For diagramatic purposes the size of particles relative to electron orbits has been much exaggerated. In real atoms there is far more empty space.)

by stage accretion known as the alpha process. Intermediate elements requiring less than the addition of a whole alpha particle (helium nucleus) are evolved in a subtle process of alpha accretion or neutron capture and subsequent radioactive decay.

It will be seen that in this way even the heaviest elements such as uranium 238 can be—and have been—built up from various combinations of the basic hydrogen building block. In stable isotopes of light and medium weight elements there are almost exactly equal numbers of protons and neutrons in the nucleus, but in heavier atoms neutrons gradually predominate until in uranium 238, the heaviest naturally occurring element, neutrons outnumber protons by a factor of more than fifty percent (92 protons and 146 neutrons—hence atomic weight 238). The number of electrons will correspond to the number of protons in the nucleus for reasons already stated.

It may sometimes happen that an electron is dislodged from its place in the outer electron shell, thus leaving the normally neutral atom with an excess of positive charge from the protons in the nucleus; or a wandering electron will take up a vacant place in atoms which have this shell incomplete, giving an excess negative charge. The resulting positively or negatively charged atom is known as an ion, and behaves in some ways like a tiny magnet. This is sometimes important in forming compounds with other elements, as the atom can now take part in a powerful form of electrostatic interaction. The atoms of certain elements (usually non-metals) are able to 'steal' electrons from the outer shells of other atoms (usually metals) where these electrons are less tightly held. The resulting molecular bonds are known as ionic bonds as opposed to the *sharing* of electrons by atoms which have a vacancy in their outer electron shells, which form covalent bonds. These bonds are also electrostatic in nature, and may be regarded as a less extreme form of ionic bond in which two 'ions' struggle for possession of pairs of shared electrons.

The molecule of common salt, in which deadly poisonous chlorine and sodium are combined to form a harmless, indeed essential compound, is an example of an ionic bond (negative chlorine ion bonds to positive sodium). The double atom hydrogen molecule H_2 (hydrogen, you will remember, has only a single electron in a shell capable of holding two) is an example of a covalent bond. Other types of bond exist, but all depend on the same principle: the attraction of unlike charges.

Very heavy atoms with a super-abundance of neutrons tend to be unstable and some, like uranium and radium, spontaneously break down into more stable forms by a process known as radioactive decay. This breaking down is accompanied by the emission of alpha, beta, and gamma radiation from the nucleus. A more drastic form of radioactive decay called nuclear fission can only occur in certain ultra-heavy unstable elements such as uranium, thorium and plutonium. In addition to the emission of gamma radiation, fission also involves the ejection of surplus neutrons at high velocities as the nucleus breaks into two main fragments. These fragments are nuclei of lighter and more stable elements such as strontium, barium, iodine and caesium which are usually radioactive and achieve greater stability by normal radioactive decay.

Alpha radiation, as the name implies, involves the emission of alpha particles (helium nuclei) in a sort of reverse action of the alpha process which builds the elements up in the first place. Alpha particles are not very penetrating and can be stopped by a thin sheet of paper. The beta particles which form beta radiation are high speed electrons, and although slightly more penetrating than alpha particles can be stopped by a sheet of thick cardboard. Gamma radiation consists of gamma-ray photons, 'energy packets' of electromagnetic waves from the high energy, short wavelength, blue end of the spectrum beyond ultraviolet and x-rays, and accompanies fission and the emission of alpha and beta particles from the nucleus. Gamma rays are very penetrating indeed, and can penetrate several inches of steel.

All these radiations disrupt and ionise the organic molecules of living tissue, causing adverse chemical changes. This makes them very dangerous to life. Fast neutrons are also very penetrating, and being nearly two thousand times more massive than beta particles (electrons) they are in effect tiny high velocity bullets. A few will probably not do much damage to human flesh because of their relatively small size, but a large number of hits would be fatal. They are rather like explosive bullets, as even after they have been brought to rest they spontaneously decay a few minutes later, emitting a beta particle (electron) and an anti-neutrino. They can also stimulate the emission of alpha and gamma radiation from atoms with which they collide. The opposite reaction, neutron synthesis, which occurs in

RADIOACTIVE DECAY

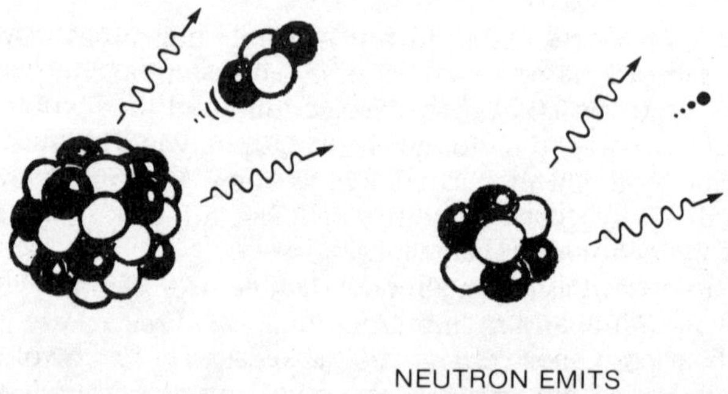

UNSTABLE NUCLEUS EMITS ALPHA PARTICLE TO BECOME NUCLEUS OF LIGHTER ELEMENT.

ALPHA DECAY

NEUTRON EMITS BETA PARTICLE TO BECOME PROTON (BETA DECAY OF CARBON 14). CARBON (6 PROTONS) BECOMES NITROGEN (7 PROTONS).

BETA DECAY

MOST HEAVY UNSTABLE ELEMENTS ACHIEVE STABLE FORM BY A COMBINATION OF ALPHA AND BETA DECAY.

the sun, is accompanied by the emission of positrons and neutrinos—another example of the symmetry of nature.

The neutrino and its corresponding anti-particle are very odd fellows indeed, having spin but no mass or charge and travelling at the speed of light. They form the most penetrating radiation known to science, and are able to pass through the entire Earth as though it were a piece of tissue paper. Thousands pass through your body every day, but fortunately they seldom interact with matter, which means that they do no damage. It also makes them very difficult to detect.

Stranger still, it has recently been discovered that there are at least three kinds of neutrinos in addition to their anti-particles and they are all believed to 'oscillate' or change form, becoming other particles in mid flight, which makes them even harder to detect than would otherwise be the case. Imagine the consternation of ornithologists trying to keep track of sparrows which could become greenfinches or linnets as they flew! There are also growing suspicions that the neutrino has a small but measurable mass which is not quite zero, and in 1980 Russian scientists claimed to have measured it. This remains to be confirmed. The last word on the neutrino has not yet been said by a long way, so this is a particle worth watching for future developments.

The fusion processes which power the sun give off great numbers of neutrinos, and these fulfil the useful purpose of instantly carrying away energy which might otherwise build up to explosive proportions in the sun's interior. (Photons, however energetic, take millions of years of absorptions and emissions to travel from core to surface.) Like all other particles the creation of a neutrino is accompanied by simultaneous production of its counterpart, in this case an anti-neutrino. The anti-neutrinos are bottled away (as mass-energy) when protons are converted into neutrons, to be released at some later date if the neutron should ever be set free from the nucleus or otherwise decay.

This is why great numbers of anti-neutrinos are constantly escaping from nuclear fission reactors, which work by bombarding the heavy atoms of uranium or plutonium with free neutrons, thus causing them to split into lighter elements and emit more neutrons. Although spontaneous fission occurs in some ultra-heavy elements, it is more usually induced when these free neutrons strike another nucleus, releasing still more neutrons and so on, causing a chain reaction. The frequency of fission depends on the likelihood of a fissionable nucleus stopping a suitable neutron and hence on the number of fissionable nuclei in close proximity.

When sufficient numbers of densely packed nuclei are involved a self-sustaining nuclear chain reaction becomes possible, but when more sparingly distributed so many neutrons are lost that the chain reaction fizzles out. For this reason fission is rare in natural uranium, in which the strongly fissile component uranium 235 is diluted with the far less active U238 and various natural impurities. In this situation alpha and beta decay (by means of which non-fissile elements such as radium also break down) becomes the main method of achieving stable form, and uranium decays through various intermediate stages to become lead.

NEUTRON SYNTHESIS AND DECAY: A SYMMETRY

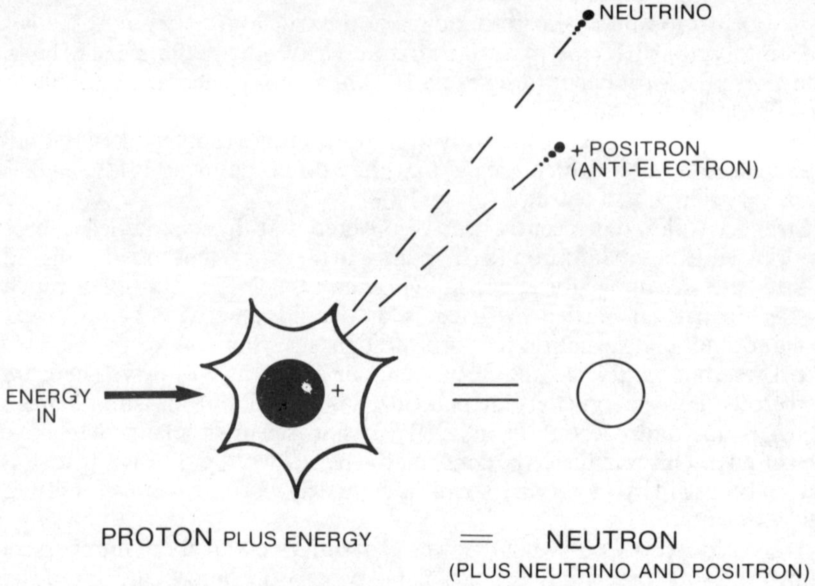

Fig. 1. Proton (1836 electron masses) absorbs energy to become neutron (1839 electron masses), emitting positron and neutrino as in sun.

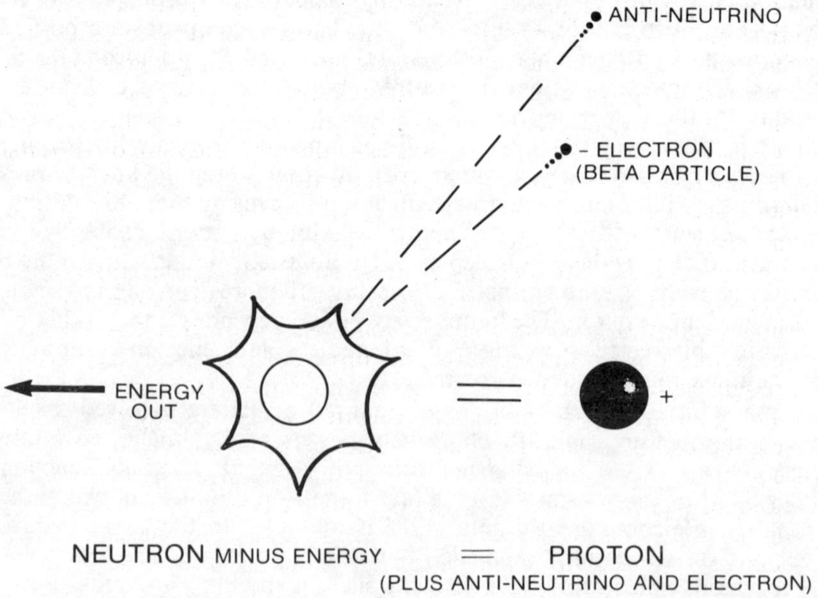

Fig. 2. Neutron (1839 electron masses) loses energy to become proton (1836 electron masses), emitting electron and anti-neutrino as in fission reactor and radioactive decay.

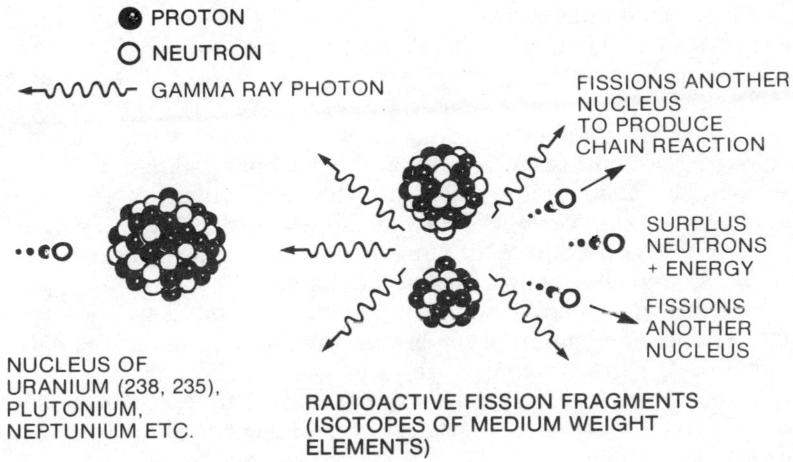

NUCLEAR FISSION

The amount of fissionable material needed to produce a self-sustaining chain reaction is called a 'critical mass'. This varies according to element, purity etc., but in atomic bombs can sometimes be only a few kilogrammes—about the size of an orange! Many fission fragments take the form of unstable isotopes with a preponderance of neutrons and are themselves radioactive, forming the dangerous wastes which are a feature of fission power stations (fusion reactors would not have these wastes and would be very much safer).

The neutrons cannot penetrate the reactor shield and are safely contained, but the anti-neutrinos emitted when they decay streak through several inches of steel, several feet of concrete, and anything else which gets in the way, to go zooming off at 186,000 miles per second. Anti-neutrinos emitted by nuclear reactors in Australia regularly pass through the centre of the Earth and through the bodies of any Englishmen who happen to be in the way, before flying off into space without the slightest loss of energy or speed. Their loss represents wasted energy to the reactor, but how do you contain particles which can theoretically penetrate eleven light years of solid lead?

The rate of radioactive decay of the various unstable atoms which, by emitting particles, break down into more stable elements is measured by the amount of time taken for half the element's atoms to decay. The strange thing is that there is no way of telling exactly when this decay will occur in any individual atom. It may be in ten seconds, in a thousand years, or in ten million years time. All that can be said with any certainty is that within a certain characteristic period of time, known as the half-life, half the atoms in a given sample of the element will undergo radioactive decay. The half-life of radium, for example, is only 1,600 years, from which you might deduce that it is more radioactive than uranium 238, the major component of natural uranium, which has a half-life of four and a half billion years.

The isotope uranium 235, the other component of natural uranium, (about 0·7 percent) is much more active, having a half-life of 710 million years. It also fissions more readily, which makes it much more suitable for use in bombs and reactors. The two isotopes are difficult to separate so in practice the proportion of the more fissile component is increased, and the resulting product is known as enriched uranium.

In a further half-life period half the remaining atoms will decay, and in the same period again another half, and so on, with the number of remaining atoms being halved each time. Although this can go on almost indefinitely with the number of atoms remaining never actually reduced to zero, it will obviously take only a few half-lives before the radiation level becomes negligible.

This combination between what seem to be arbitrary natural laws on the one hand and random chance on the other is seen many times in nature, and helps to produce the almost poetic effect of order and harmony which will continue to fascinate scientists and thinkers as long as the human race exists. The elegant logic and symmetry typified by particle anti-particle materialisation and annihilation is yet another facet of the aesthetic appeal of truth to the higher aspects of the human mind.

Another example of randomness in nature concerns the penetration of matter by neutrinos and anti-neutrinos. Although theoretically able to penetrate a wall of lead billions of miles thick, a small proportion will be stopped at random intervals inside this distance by collision and interaction with atomic nuclei. An infinitesimally minute proportion will even be stopped in the first millimetre of lead, but there is no way of telling in advance which ones will stop here, which will be stopped a yard or a million miles further on, or which will penetrate clean through. It is only the fact that a tiny minority will be stopped before they have gone very far which allows us to detect them at all.

Before leaving this brief survey of the atom I will also mention the meson, which plays an important role as the particle of the strong interaction which binds together the protons and neutrons in the nucleus. The meson is a particle of intermediate mass (some 270 times the mass of the electron) which is rapidly shuttled between particles in the nucleus and exerts a powerful force which binds them together. Its role may be loosely compared to a ball between two players which is repeatedly grabbed first by one and then the other, setting up an attractive force. In the atom this force has to be strong enough to overcome the mutual repulsion of the positively charged protons in the nucleus. An ephemeral particle continually created and re-created from borrowed energy, the meson is unstable outside the nucleus and decays in a minute fraction of a second, forming two gamma-ray photons.

Now back to the beginning again and the 'Big Bang'—the mightiest explosion there has ever been. A torrent of radiant energy from another dimension in space and time expanded outwards in all directions at the speed of light. Initially the explosion would have consisted mainly of high energy gamma-ray photons, and within a few seconds materialisation had begun as electron anti-electron, proton anti-proton pairs appeared, to take the place of vanishing gamma rays. Then protons began to collide, absorbing energy, and neutrons began to appear. After a few minutes expansion had sufficiently cooled the white-hot mass for hydrogen isotopes and atoms of helium to form.

Things were happening very quickly at this stage in the birth of our universe. Owing to rapid expansion and cooling very little synthesis of elements took place beyond this stage, so that the early universe was composed almost entirely of hydrogen and helium, and to an only slightly lesser extent still is.

We now come to a strange paradox for which there is so far no really convincing explanation. It might be called 'the case of the missing anti-matter'. There is no known mechanism which creates matter without creating an equal quantity of anti-matter. As we have seen, gamma ray materialisation or pair production always produces two particles: creation of an electron is always accompanied by simultaneous creation of a positron, or in the case of a proton the creation of an anti-proton. In the first few minutes of the birth of the universe there must have been equal quantities of both matter and anti-matter. What has happened since to all the anti-matter?

The world we see around us today consists almost entirely of matter; anti-matter is extremely rare and normally found only in the form of short lived sub-atomic particles. This is just as well, as any large quantities of anti-matter would need to be kept permanently isolated in a vacuum, perhaps levitated by means of magnets.

Anti-matter looks exactly the same as matter and behaves in the same way chemically. Anti-hydrogen would burn merrily with anti-oxygen, and the resulting anti-water would look, weigh, and behave exactly the same as odinary water—until you poured it into a mug or beaker made of matter. There would then be an almighty explosion equivalent to the detonation of a multi-megaton hydrogen bomb as particle and anti-particle annihilated each other to produce pure energy!

One solution to the case of the missing anti-matter which has been suggested is that somehow it coalesced to form separate galaxies. These galaxies would look exactly the same as 'ordinary' galaxies. There is no proof at the moment one way or the other, as the light which left them would be ordinary light. So there may be anti-galaxies with anti-stars around which circle anti-planets complete with anti-life forms. It is worth emphasising that any intelligent being on one of these planets would see himself and his surroundings as perfectly normal, and he may at this very moment be speculating as to whether *our* galaxy (the Milky Way) is made of anti-matter. Which is which depends on your point of view—one is just as valid as the other.

The great weakness of this theory at present is that no one has yet proposed a mechanism whereby matter and anti-matter might divide themselves into separate galaxies. One way the problem might be resolved is by the development of a suitably sensitive neutrino telescope. The sun and most other stars emit huge quantities of neutrinos as a consequence of their thermonuclear fusion processes. It follows that these same reactions taking place in an anti-star in our hypothetical anti-galaxy would emit anti-neutrinos rather than neutrinos: therefore a stream of anti-neutrinos emerging from a distant galaxy would indicate it to be made of anti-matter. This would be an extremely difficult proposition, however, as neutrinos (or anti-neutrinos) are almost impossible to detect, especially with the kind of sensitivity this experiment would require.

A primitive neutrino telescope has already been built one mile deep in an American gold mine (to shield it from irrelevant radiations such as cosmic rays), and this has been observing the neutrino flux from the sun for several years. Results seem to indicate that conditions at the centre of the sun, or neutrinos, or perhaps both, are less well understood than we had thought. A more sophisticated instrument may one day be used to observe the galaxies. Even a negative result indicating no anti-matter galaxies would be of some value, as we would then know that we must look for an answer somewhere else. Perhaps some quite different and as yet unsuspected solution may be found, but at present the problem remains one of nature's greatest mysteries.

Back at the rapidly cooling fireball of the Big Bang, matter and anti-matter had somehow sorted themselves out. The primeval fireball dispersed and cooled as it scattered hydrogen in all directions. The myriads of filaments and eddies from the explosion began to condense into great globes of hydrogen gas which drew away from one another at thousands of miles per second. The universe grew dark as it expanded, but gradually the smaller irregularities within these galactic spheres began to heat up as they were crushed and compressed by gravity, and soon the universe began to glow again as, like millions of cosmic fireflies, the first stars began to form.

Initially this incandescence was caused solely by gravitational collapse, but after a few thousand years the temperatures and densities at the cores of these first generation stars became sufficient for hydrogen fusion to commence. Until this time the universe consisted almost entirely of hydrogen with a few traces of other elements (mainly helium), but now billions of stellar furnaces began to consume this hydrogen and synthesise the heavier elements.

Small stars like our own sun have a relatively modest energy output and can power themselves on hydrogen fusion for about ten billion years before their central regions become so choked with helium that hydrogen fusion ceases. The core of helium 'ash', deprived of support by radiation pressure, contracts and heats to even higher temperatures. The heat and pressure are then so high that fusion of helium into beryllium and carbon commences, the first stage in the alpha process which helps build up many of the lighter elements below the weight of iron. Beryllium is unstable in the conditions believed to exist at the centres of stars and there is some

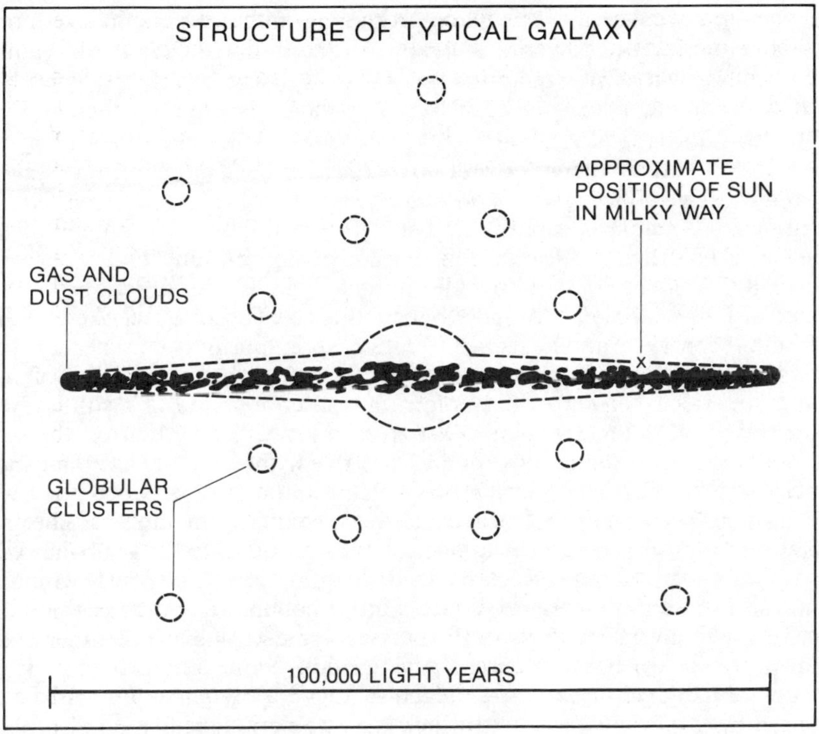

controversy as to the origin of the observed cosmic abundances.

The early proto-galaxies were spherical in shape but spinning faster as they contracted. This rotation gradually flattened the main body of the galaxy into the lens-shaped disc which we see today, but a halo of globular star clusters still delineates the former spherical outline. Just why these globular clusters were not drawn into the rotating disc of the galactic plane has not yet been satisfactorily explained, but nevertheless there they are for anybody to see. Globular clusters are made up of very old stars which are tightly packed together and show little trace of heavy elements in their outer layers, indicating that they were among the very first stars to form. A few galaxies still largely retain their original spherical shape, and these are called elliptical galaxies.

Another galactic structure which has been difficult to explain is the catherine-wheel effect produced by the spiral arms in the majority of galaxies, including our own. The current view is that this is the result of an interaction between rotating gas and the galaxy's magnetic field. It is also suspected that the arms are renewed every 100 million years or so by violent shockwaves created by shells of expanding gas flung off from explosions within the galactic nucleus. These arms are now the main birth place of stars. Eddies and local compression in the swirling hydrogen gas cloud, perhaps aided by the magnetic field, produced concentrations of gas sufficiently dense for the combined effects of atomic gravitational fields to take over and prevent dispersion. Although the weakest of the four fundamental forces at an atomic level, the gravitational interaction gains effectiveness where vast quantities of matter become sufficiently concentrated, becoming increasingly potent as it gathers the mass together.

Giant stars, typically several times more massive than the sun, are much more lavish in their energy expenditure and reach the helium burning stage in only a few million years, and subsequent stages of element synthesis continue at a similarly rapid pace. I use the word 'burning', of course, in the sense of thermonuclear fusion. Paradoxically, burning or fire as we normally think of it is impossible at such high temperatures, however abundant the supply of oxygen. When the core becomes choked with carbon, it contracts and heats again until carbon 'burning' commences. In this way the star builds up a layered structure somewhat reminiscent of an onion, in which heavier elements form at the core supporting still active outer layers of lighter elements where fusion reactions continue.

But when fusion reactions at the core reach the stage of synthesising nuclei of intermediate size such as iron, nickel and cobalt, they can proceed no further. The creation of heavier elements requires an input of energy rather than producing any. The nuclear fires go out. No longer balanced by opposing radiation pressure, the forces of gravity become overwhelming and the inert core undergoes a catastrophic collapse into which the still active outer layers must also fall. This generates colossal pressures and temperatures which cause a runaway reaction in the unburnt fuel. The energy of the gigantic explosion which follows is only partly absorbed by the synthesis of heavier elements which is now able to take place and which barely serves to delay the inevitable. The star's outer layers are ripped apart, and blasted into space in an explosion of awe inspiring dimensions. This stellar explosion is known as a supernova and the still expanding remnants of one of them, the well known Crab Nebula, can still be seen. For a few weeks such a star can outshine ten billion suns, and its expanding shell delivers great quantities of heavy elements to enrich the inter-stellar medium.

This was a very important development in the history of the early universe, as life or even planets as we know them could not form in a universe of almost pure hydrogen. All the heavier elements in your body and the things you see around you were created in the interiors of stars and set free for more useful purposes by just such explosions as I have described. Only second generation stars such as our own sun are rich in such elements and therefore suitable abodes for life.

Element synthesis fueled by the explosion continues for some time in the expanding shell by a process of neutron capture—the same process by which plutonium is built up from otherwise useless U238 in fast breeder nuclear reactors. It has been speculated that in this way super-heavy elements are formed far heavier than any now remaining on Earth, but being both scarce and unstable have long since decayed below the threshold at which they are detectable. The explosion also emits energetic nuclei at speeds of many miles per second, and these are accelerated to even higher speeds by inter-stellar magnetic fields to form dangerous cosmic rays. An object far too small to be seen in the most powerful electron microscope can have the kinetic energy of a rifle bullet which, though trillions of times larger, is very much slower.

Some cosmic rays are probably formed by other stellar processes and consist mainly of fast protons (hydrogen nuclei), but also much heavier nuclei up to and perhaps even beyond uranium. These atomic nuclei, travelling at speeds which in some cases approach that of light itself, are constantly bombarding the upper atmosphere where they collide with atoms of nitrogen or oxygen and disintegrate into showers of sub-atomic particles. These showers form dangerous secondary radiations which, fortunately for us, are usually stopped by the atmosphere before reaching the ground, but the effects are still easily measurable in high flying aircraft and to a lesser extent even at sea level. The detection of anti-matter nuclei in cosmic rays would be significant, though positrons and anti-protons can be expected to occur as a result of already familiar stellar processes.

At the core of the dying star matter is crushed almost out of existence. The atom, like the solar system, consists mostly of empty space with most of its mass concentrated in the central nucleus. But the pressures created at the heart of a supernova explosion are so great that the planetary electrons are crushed into the nucleus where they are absorbed by the protons, which are thus transformed into neutrons. The resulting stellar remnant is called a neutron star and in some respects resembles a gigantic atomic nucleus composed almost entirely of neutrons. Although only a few miles in diameter a neutron star contains about twice as much material as the sun, which is 860,000 miles in diameter but made up of atoms containing mostly empty space and a plasma of ions and sub-atomic particles with empty space between them. A spoonful of neutron star material (sometimes called neutronium) would weigh millions of tons, even in the feeble gravitational field of Earth. Neutron stars are also known as pulsars, as some if not all of them emit beams or pulses of electromagnetic waves, and the first to be discovered was detected as recently as 1966 at the heart of the Crab Nebula.

You might think that this represented the ultimate density of matter, but you would be wrong. It seems that supernovae involving stars of more than 9 or 10 solar masses can create pressures which even neutrons are unable to withstand and gravitational collapse of the core proceeds to a point at which a new type of stellar object is evolved, smaller and denser than a neutron star. These objects are called black holes and are so called because though matter can fall into them, nothing, not even light, can come out.

The speed to which a rocket, cannon ball or any other object must accelerate in order to escape the gravitational pull of Earth is 7 miles per second, and is termed escape velocity. The escape velocity of a black hole exceeds 186,000 miles per second, which is the speed of light, and this makes them very difficult to detect as not even light can escape them. However, they can be observed indirectly when they sweep up matter from their surroundings. As this material is sucked down into a gravitational vortex it is compressed and heated until it emits x-rays, and in this way at least

BLACK HOLE ACCRETION DISC

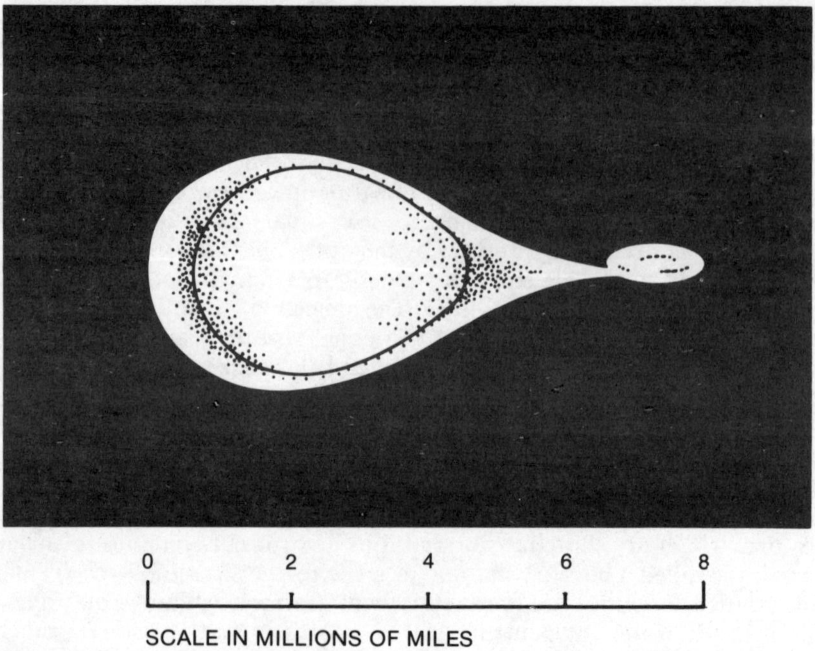

10 solar mass black hole orbiting 15 solar mass giant companion. Giant star is distorted by tidal forces and gas is drawn off to form accretion disc before spiralling into black hole. On this scale the 30 mile diameter event horizon is far too small to be seen.

one black hole has been detected—the x-ray source Cygnus X-1. I should perhaps mention that owing to the inherent elusiveness of black holes it is unlikely that any example can be identified with a hundred percent certainty, and Cygnus X-1 is no exception. However, with better than ninety percent certainty it is probably about as good as we will get, and several other likely candidates are also known.

It seems extremely unlikely that stars exceeding about 10 solar masses will always eject enough mass to avoid becoming black holes. If Einstein was right (and there is considerable evidence that he was), stellar remnants of more than 3·2 solar masses must become black holes when no longer supported by radiation or other internal pressure. Even in a Newtonian universe there would be nothing to stop the most massive stars eventually degenerating into black holes (though not space-time singularities) provided they did not expel most of their mass as a supernova. The sun is far too small to become a black hole or even a neutron star. Instead, after first swelling to become a rarefied red giant during the helium burning stage (rarefied in the outer layers only) it will finally shrink to a super-dense sphere barely larger than the Earth, called a white dwarf.

Because of compression by gravitational forces white dwarves are very hot (hence their colour) but give off very little radiation owing to their

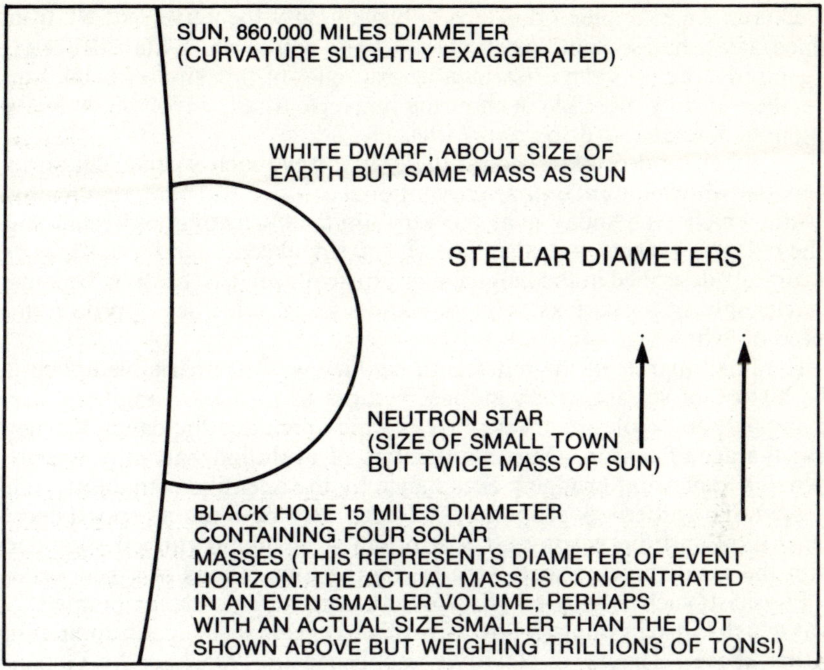

small size. As the inner planets will be first scorched then frozen, all life on Earth will cease. But there is no need to worry: all this will not happen for another five or six billion years, longer than the Earth has so far existed. Until then the sun will continue to shine in the same placid, reliable way it has always done. Though far less dense than a neutron star, a spoonful of white dwarf material would weigh several tons, composed of a plasma in which ions and subatomic particles are almost touching, leaving far less empty space than in a normal star, and the pressure of a 'degenerate' electron gas is sufficient to stave off further collapse.

Relatively small amounts of matter can theoretically become black holes when compressed to sufficient density, as this increases the intensity of their gravitational fields. There has been much speculation as to the existence of mini black holes about the size of marbles containing a mass equivalent to the entire Earth, but there is not the slightest evidence to support such ideas and even on theoretical grounds the chances are slender. Such things are probably no more than the figments of mathematicians' imaginations.

Theoretically there are two types of black holes (i.e. objects with escape velocities exceeding 186,000 miles per second). The first kind involve such large quantities of material that it can be sufficiently dispersed for matter to exist more or less in a form we can understand, and tidal forces within the 'event horizon' (a gravitational boundary from inside which no light or anything else can escape) are not sufficient to break down all organised structures. An example of this type of black hole is the universe itself, from which, as we shall see, nothing ever escapes and within which quite delicately organised structures can exist. Smaller examples of this kind of black hole are theoretically possible within the universe itself, involving massive amounts of relatively dispersed matter.

The other kind of black hole usually involves much smaller quantities of matter which has undergone gravitational collapse and compression into a state which even today is at the very limits of scientific understanding. These latter objects are also known as singularities and can only be accurately described in the language of Einstein's relativity, which becomes increasingly important as velocities and escape velocities approach the speed of light.

The final quarter of the nineteenth century was a remarkable epoch in the history of science, when the last vestiges of mediaeval thinking were giving way to revolutionary new ideas which heralded the dawn of a new era: the age of reason. Darwin's theories of evolution had only recently gained a somewhat grudging acceptance by the scientific community, and at scientific gatherings proponents of such archaic ideas as spontaneous generation and the Adam and Eve origin of mankind rubbed shoulders with the great thinkers who were laying the foundations of a new order.

Physicists such as Maxwell, FitzGerald, Lorentz, and Planck pointed the way which Einstein was to follow so brilliantly a few years later, culminating in his Special Theory of Relativity published in 1905 and the General Theory which followed it in 1915. It is amazing to think that ideas which even today are considered advanced had their origins well back in the nineteenth century when many scientists still believed in magic!

The essence of Einstein's theory was that there is no such thing as absolute motion or absolute rest. The only motion which has any validity is motion with respect to some other object or objects which form a frame of reference. For example, you are now at rest relative to someone sitting in the same room, but relative to an observer at the North Pole you would be streaking along with the rotation of the Earth at about 600 m.p.h. To another observer in the same frame of reference as himself the man at the North Pole would appear to be at rest, but to an observer at rest with respect to the sun they both would be travelling at 18 miles per second as the Earth moved in orbit. The entire solar system is circling the galaxy at nearly 200 miles per second. Properties such as length in space or duration in time appear to change when viewed from different frames of reference.

This change is so small for objects in relative motion at comparatively low speeds that it is often unmeasurable and can be ignored, but for things moving at a significant fraction of the speed of light the changes are substantial and Newtonian physics will no longer describe them accurately. The one constant which never varies and is the same in all frames of reference is the speed of light, which travels at 186,000 miles per second. Therefore light emitted from a galaxy receding from us at 100,000 miles per second will not reach us at a relative speed of 86,000 miles per second as you might expect, but at the full normal speed, in contrast to high speed particles of matter which would arrive at less than their starting velocity (though not as much less as simple arithmetic suggests). The light loses energy, however, by the lengthening of its wavelength towards the red end of the spectrum, thus producing the red-shift observed by astronomers. This invariance of the speed of light, together with the abandonment of absolute rest or motion, is one of the cornerstones of Special Relativity.

It might take you three seconds to read this sentence, but to an observer in a frame of reference moving at velocities approaching 186,000 miles per second relative to your own it might seem that you took three minutes. To such an observer passing you in a vertical direction you would appear to be no taller than a fraction of an inch. However much you might protest that you were nearly six feet tall and had read the sentence in three seconds it would seem to him that your clocks were running slow and your yardsticks and measuring instruments had shrunk. He and his yardsticks in their turn would seem to have shrunk when viewed from *your* frame of reference, and *his* clocks would be running slow.

Time is change. If nothing ever changed there would be no time. Whether that change is fast or slow depends on the frame of reference used for comparison, just as the question of whether an object is large or small depends on such a frame of reference and is meaningless in itself. An amoeba is small compared with a man but gigantic compared with a virus. It makes no sense to state that an amoeba is large or small unless the frame of reference is also understood. The question of whether observer A's time (rate of change) is fast or slow is meaningless in itself unless comparison can be made with a suitable frame of reference.

If you and your environment changed at half their present rate (in other words, if your time dilated) you would know nothing at all about it! The slowing would only be detectable by comparison with a reference frame which had not changed and which would therefore seem to have speeded up. Where time is concerned there is no absolute 'fast' or 'slow', just as 'large' or 'small' are not absolute qualities but relationships between one object and another.

Sir Arthur Eddington gave a similar common-sense explanation when asked whether objects 'really' contracted in the direction of motion. He said that length was not an absolute quantity, entire of itself, but a relationship between an object and an observer. Length was meaningless unless the observer was also specified. The energy of motion (kinetic energy) is another obvious example where magnitude is dependent on the relationship with an observer. Even the *direction* of time is dependent on this relationship. At faster-than-light speeds cause would no longer precede effect and the flow of time would be reversed, a logical absurdity which is no doubt the reason why nature forbids superluminal travel (at least in this universe!).

So the question of whose clocks are really slow, like the question of whether light consists of particles or waves, is for practical purposes not very meaningful. The important thing is that when measuring anything in motion at very high speeds relative to ourselves, these distortions must be taken into account if we require perfect accuracy. A further consequence of relativity is that mass and energy are one and the same; therefore anything in motion relative to ourselves has not only additional energy but additional mass.

For example, two particles, each travelling at 99 percent of the speed of light and converging on a target from opposite directions, will not be approaching each other at nearly twice the speed of light (as common-sense demands) but at a speed which falls short of the speed of light by the merest fraction of a percent! The missing velocity, however, is manifested as the increased mass of the particles. Similarly, two galaxies receding from Earth at 75 percent of the speed of light, but in precisely opposite directions, are not receding from each other at 150 percent of the speed of light as might first appear to be the case, but at less than 95 percent. Relativistically it may seem at first glance that two and two sometimes equals three, but once we take mass-energy into account they equal four again.

This increase in mass is unmeasurable in the case of a motorcar or an aeroplane, but in the world of the atom where particles often travel at speeds approaching that of light the increase in mass and other relativistic effects are very significant indeed.

If we try to accelerate sub-atomic particles in an accelerator to the speed of light, the faster we accelerate them the more they increase in mass but, because of inertia, the more their mass increases the harder they are to accelerate. When speed is about 99·9 percent of the speed of light we find that nearly all the magnetic energy we pump in to accelerate them is manifested as an increase in mass, and as massive objects are much harder to accelerate than light ones it becomes almost impossible to accelerate them any further however much energy we pump into them. And this in spite of the fact that by the calculations of Newtonian physics they have received enough energy to exceed the speed of light many times over! It is worth stressing that the effects are reciprocal and to an observer travelling with the particles the particles themselves would appear unchanged. What would appear to change, however, is the length of the accelerator, which at 99·9 percent of the speed of light would (from the point of view of the particles) have shrunk almost to zero, leaving insufficient time and space for further acceleration.

The symmetry is always maintained. From the particles' point of view it is the mass of the *Earth* which would seem to have changed as it flashed by at almost the speed of light, and would appear to have increased many times over. More peculiar still, an input of energy to the particles which is manifested to one observer (who sees himself at rest) as an increase in mass will be seen by an observer in uniform motion at a similar speed and direction to the particles as an increase in velocity. The same energy can manifest itself as mass to one observer and velocity to another! (In a similar way photons may be infra-red to one observer, but to another observer in opposite and relativistic motion they will be blue-shifted to gamma wavelengths and able to materialise into electrons. This actually happens in outer space, where atomic nuclei—cosmic rays—routinely collide with and materialise low energy photons not normally capable of materialisation.)

It is significant that to an observer travelling with the particles neither he nor the particles nor anything else which even approximately shared their motion would seem in any way odd. This appearance of normality is shown in your own surroundings, which seem to you to be perfectly normal but would manifest all manner of relativistic effects to an astronomer in some distant galaxy (many of which are known to be receding from us at more than half the speed of light). It is equally valid to say that we are receding from them, as in reality we do not enjoy some special position at the centre of the universe where we are at rest with all the other galaxies moving away from us. An astronomer in any other galaxy would see his own as being at rest while others receded in all directions.

The best way to understand this is to take a spherical balloon and mark out a number of evenly spaced dots on its surface. When the balloon is blown up each dot will be surrounded by others moving away from it, with more distant dots separating at faster speeds than dots which are closer together. An ant sitting on any dot would see himself as being at rest while all the others moved away, with the furthest dots moving away fastest of all. The balloon represents the expanding universe, the dots are the receding galaxies and you are the ant. You do not feel yourself to be moving at half the speed of light any more than you feel yourself to be orbiting the sun at 18 miles per second or the Milky Way at nearly 200 miles per second, but from the frame of reference of an astronomer in a distant galaxy you are. Which one of you is really moving is just another of those irrelevant questions already referred to (but in this case the answer is: both).

The experiment with the accelerated particles has actually been done with electrons in an American linear particle accelerator (at Stanford University in California), where it was found impossible for the electrons to attain the speed of light no matter how much energy they were given, thus demonstrating the validity of relativity theory. In a frame of reference where an object is moving at the speed of light relative to ourselves its mass becomes infinite, its clocks and therefore its time (but not ours) stand still, and its length in the direction of motion becomes zero, like a reflection in a mirror when viewed edge on.

For those who feel that all this defies common-sense there is just one small grain of hope. Einstein himself felt that his theory was not perfect, and would one day need to be modified and for some purposes superseded by what he called a unified field theory, just as Newtonian theory had been by relativity. He spent the latter years of his life searching in vain for this unified field theory, which would explain rather than merely describe gravitation and other force fields in a single all-embracing theory. The world is now waiting for a new Einstein or Newton to meet the challenge and formulate a unified field theory which, if it ever comes, may indicate small imperfections and inaccuracies in relativity. There are also areas of conflict between relativity and quantum theory which the new theory would hopefully resolve.

The fact remains that both the Special Theory of 1905, which examined the effects of uniform motion, and the General Theory of Relativity, which followed it ten years later and examined the effects of accelerated motion and gravitation, have been subjected to many experimental tests and so far passed them all with flying colours. Einstein's theory will never be discredited any more than was Newton's, but just possibly it may one day need to be slightly modified. The snag is that far from simplifying everything the overall effect of a unified field theory may be to reveal a universe or universes even more astounding than that indicated by relativity.

To return to the subject of black holes, relativity theory seems to indicate that once gravitational collapse has proceeded beyond a certain point it can no longer be stopped or even resisted. The material in the star is crushed out of existence, perhaps to appear as a white hole in some other universe, leaving behind only its gravitational field like the grin of the Cheshire Cat!

Dark objects such as black holes are of vital importance to the future of the universe, which at present continues to expand. The most distant galaxies yet observed are about 12 billion light years away, almost at the edge of the observable universe. This means that we are seeing them not as they are now but as they were 12 billion years ago when the light left them. In the same way we see the moon as it was $1\frac{1}{4}$ seconds ago, the sun as it was nearly $8\frac{1}{2}$ minutes ago, and the nearest star as it was about $4\frac{1}{2}$ years ago. Therefore any astronomers in these most distant galaxies, which are receding from us at nine tenths of the speed of light, are seeing the Milky Way (our own galaxy) not as it is now but as it was 12 billion years ago. It follows that they cannot see our sun, however good their instruments, as it was formed only five billion years ago and its first rays have travelled less than half way to their telescopes.

The further our telescopes reach out into space the further we see back in time, and this raises an exciting prospect: if the universe is 15 billion years old and we can already see light which left it 12 billion years ago, then we are seeing events which took place 'only' three billion years after the explosion of the primeval fireball. Perhaps if we had bigger and better telescopes able to probe deeper into space we could look back at the very birth of our universe—at the Big Bang itself! Unfortunately there appears to be an insurmountable barrier. The further away a light source is the more its radiations are weakened and attenuated by the expansion of the universe. The wavelengths are drawn out and reddened by the Doppler effect which causes the red-shift. At 12 billion light years the speed of recession is already nine tenths of the speed of light, and at the speed of light itself (or as near to it as material objects can get) the wavelengths become so drawn out that they are impossible to detect—they have red-shifted to infinity. This marks the boundary of the observable universe which cannot be penetrated however large our telescopes or however perfect our instruments.

If present estimates of the age of the universe are correct it seems that direct observation of the Big Bang is forever tantalisingly just beyond our reach. However, indirect evidence of the fireball has been detected by radio telescopes which have picked up micro-wave background radiation corresponding to a temperature of 3°K, evenly distributed in all directions. This radiation, which seems to permeate the whole of space, is thought to be the attenuated remnant of the flash of the primeval fireball, and agrees well with theoretical calculations.

It was predicted by Einstein in 1915 and subsequently verified by experiment that light does not travel in straight lines but is bent by gravitational fields. Light and other electromagnetic radiations, together with neutrinos and anti-neutrinos which travel at the same speed, do not leave the universe but are deflected by the enormous gravitational field, presumably orbiting the universe until they meet and interact with other particles. The universe is therefore a closed system from within which nothing can escape. The same powerful gravitation which bends the paths of light rays and neutrinos is clutching at the speeding galaxies and slowing them down. Like debris rising from an explosion the galaxies are gradually brought by gravity to a standstill, at which point they will slowly fall back upon themselves as the explosion spends itself and collapses inwards. This point will not be reached until about 40 billion years hence, long after the sun has burned itself out and all life ceased to exist on Earth.

Some calculations of the amount of material in the universe indicate that there is not enough to produce the required gravitational field and therefore the universe will go on expanding for ever, but these estimates fail to take into account the amount of dark and almost invisible matter in the universe. These dark objects include infra-red stars and brown dwarves (stars of low mass whose nuclear furnaces were unable to get started), stars hidden in the dark clouds or nebulae which pervade the galaxies, black holes, and many faint objects such as neutron stars, white dwarves, and red dwarves which are not visible unless they are relatively close to us. In addition, it has been discovered that vast amounts of invisible matter amounting to ten times the mass of the visible component surround many of the galaxies. New matter is being discovered almost annually.

Black holes alone can be very massive and may account for enormous quantities of matter. If the number of stars large enough to end their days as black holes is any guide, they may be relatively common objects. There is evidence to suggest that the strange objects emitting prodigious amounts of energy at the centres of most galaxies may be super-massive black holes of equivalent mass to millions of suns! There is even one at the centre of our own galaxy which is, however, a bit of a damp squib compared to some found elsewhere. The energy produced by quasars and Seyfert galaxies cannot be accounted for by ordinary stellar processes, and if some matter anti-matter reaction is not involved this is best explained by the existence of gigantic black holes, though white holes have also been suggested.

One very speculative suggestion to account for the observed energy production is this: as we are seeing events which took place when the universe was very young (the light from quasars has taken billions of years to get here), are we witnessing the disposal of vast quantities of anti-matter? It is probably significant that the activity of this energy source declines with age and is much more subdued in evolved and nearby galaxies. What is known with some certainty is that these energetic centres are of very small diameter, probably much less than one light year, and situated at the very heart of the galaxy, which makes them difficult to observe. This is consistent with the existence of large black holes, and there is nothing we know at present to rule out the possibility that they may be made of

THE PULSATING UNIVERSE *(see diagram opposite)*

Unlike explosions on a small scale, where light traverses the explosion instantly in a straight line and carries information to external observers, in this explosion light moves with almost infinite slowness between the fragments of debris and dramatically loses energy with expansion. Its path is curved. Because of this curvature and loss of energy, it is trapped within the explosion and unable to carry away information or energy to regions beyond the event horizon. This presents a very strange picture of the universe to observers, quite unlike what they would expect to see by analogy with conventional explosions where forces are insufficient to bring relativity into play. In these small scale explosions there is no 'event horizon', no bending of light or red-shift or blue-shift of any consequence, and information transmitted from events in all regions of the expanding material reaches an observer at the very instant of their occurrence.

anti-matter!

In certain circumstances black holes can be very efficient producers of energy. When sufficient matter is swept up from their surroundings it spirals in on an accretion disc, compressed and heated as it goes, but before disappearing for ever beyond the event horizon it emits frantic distress calls in the form of energetic electromagnetic radiation.

Inter-stellar and inter-galactic space is not a perfect vacuum. Within the galaxy even the most transparent regions have a hydrogen atom every few metres and for several light years around many stars a dense solar wind of speeding protons and electrons streams incessantly into space. Trillions of tons of energy in the form of photons, neutrinos and cosmic rays are streaking between the galaxies and perhaps orbiting the universe, so if energy is a form of mass this presumably affects the gravitational field. In addition, huge inter-galactic clouds of hydrogen gas many light years in diameter have recently been detected by radio telescopes. It seems certain that the amount of visible matter in the universe is far exceeded by the amount of dark or near invisible matter, and this is borne out by more indirect methods of estimating the mass of the universe.

The film of the universe is now run backwards as the galaxies fall back and converge towards a mutual centre of gravity. Instead of the previous universal red-shift we now have a universal blue-shift which before long produces a strange reversal of Olbers' paradox. Olbers was a nineteenth century astronomer who reasoned that if the universe went on for ever a line of sight in any direction should eventually reach the surface of a star. Why then was the sky so dark when theoretically it should be ablaze by night as well as by day? Although the universe is finite and does not continue for ever, it is sufficiently large for the paradox to be real. The problem is resolved when we take into account the expansion of the universe and the consequent red-shift which weakens and attenuates the light rays until beyond a certain distance (the boundary of the observable universe) they are no longer visible. Another factor is the large amount of dark obscuring matter in the galactic nebulae. This process is reversed as the galaxies rush together, as the blue-shift intensifies rather than weakens the light rays.

Above: The spiral galaxy in Andromeda. Over two million light years from Earth, this is the nearest star system similar to our own (though half as large again as the Milky Way).

Below: NGC 4565 in Coma Berenices is a spiral galaxy seen edge on, and is similar in appearance to a panoramic view of our Milky Way.

Above: The nebula in Orion is a great cloud of gas and dust forming only a small part of our Milky Way. In this and similar gas clouds, new stars are constantly being born.

Below: The Crab Nebula, remains of an exploding star or supernova observed by astronomers in 1054 AD. In 1966 the first neutron star to be discovered was detected at its core.

Above left: Electron micrograph of gold atoms in crystal lattice (× 12,500,000). In the early years of the 20th century there were still several prominent scientists who did not believe in the physical reality of atoms!

Above right: Cosmic-ray particle travelling at almost the speed of light collides with proton in bubble-chamber. Despite relativistic velocities and creation of many new particles, the total mass-energy of particles leaving the reaction must exactly equal the mass-energy which went in. Neutral particles leave no tracks.

Below: Bubble-chamber photograph showing three cases of matter/anti-matter materialisation. Gamma-ray photons (leaving no tracks) enter chamber from left to right. On encountering protons in liquid hydrogen they materialise into electron/positron pairs which leave wakes of tiny bubbles. A magnetic field bends negatively charged electrons toward the top of the picture while the oppositely charged positrons leave tracks which curve downwards (tiny crosses are simply calibration marks and can be ignored).

Above: Albert Einstein in his later years. His most important work was done before the age of forty.

Below: Apollo moon rocket leaving the launch pad. After burning some 2,600 tons of fuel the third stage finally attained a velocity of 7 miles per second and a relativistic mass increase of about a thousandth of a gramme!

THE PULSATING UNIVERSE

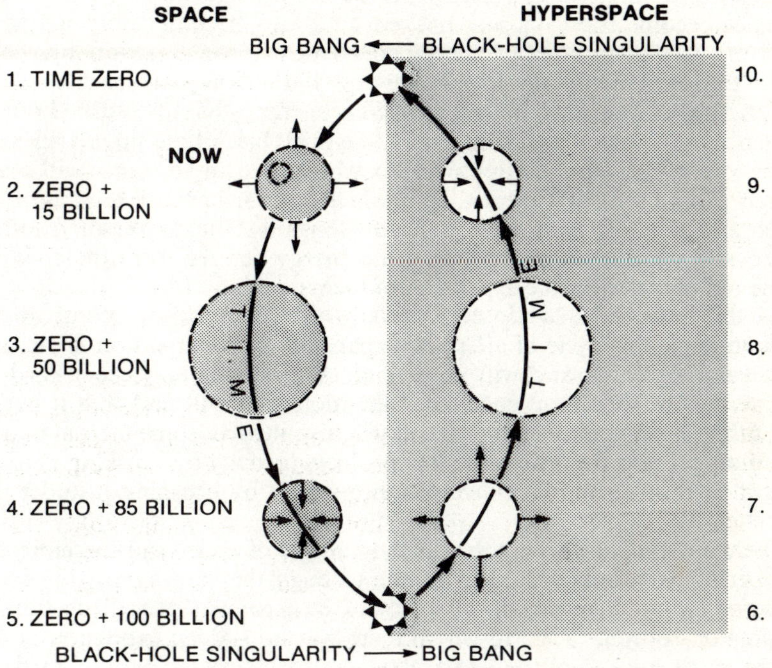

1. BIG BANG, A SUPER-RADIANT WHITE HOLE FROM WHICH PRESENT UNIVERSE EMERGES.
2. EXPANDING RED-SHIFTED UNIVERSE NOW (SMALL CIRCLE = THE OBSERVABLE UNIVERSE).
3. MAXIMUM LIMIT OF EXPANSION. UNIVERSE BEGINS TO COLLAPSE.
4. BLUE-SHIFTED COLLAPSING UNIVERSE HURTLES TOWARD DESTRUCTION.
5. UNIVERSE COLLAPSES INTO SUPER-MASSIVE BLACK HOLE SINGULARITY.
6. BIG BANG: UNIVERSE EMERGES INTO ANOTHER DIMENSION. ARROW OF TIME REVERSED.
7. EXPANDING RED-SHIFTED UNIVERSE (ALTHOUGH TIME REVERSED, CAUSE STILL PRECEDES EFFECT).
8. MAXIMUM LIMIT OF EXPANSION. UNIVERSE BEGINS TO COLLAPSE.
9. BLUE-SHIFTED COLLAPSING UNIVERSE HURTLES TOWARD DESTRUCTION.
10. UNIVERSE COLLAPSES INTO SINGULARITY. THE CYCLE REPEATS *AD INFINITUM*.

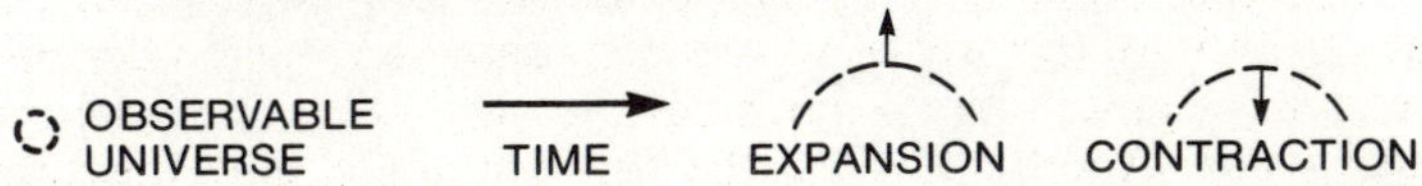

This will gradually brighten the night sky until eventually it blazes bright as day.

Gravity reasserts the crushing grip from which the galaxies had almost broken free as the collapsing universe hurtles towards destruction. As the galaxies become more closely packed collisions become more frequent, releasing vast quantities of energy to create a reverse action playback of the primeval fireball. Finally, in perhaps 100 billion years from now, the contracting universe will be reconverted into energy as it is crushed out of existence in the black hole singularity to end all black hole singularities... drawn down into a gravitational vortex where time and distance will cease to exist, to be violently ejected as a white hole in another dimension in space and time. The universe is dead—long live the universe! Another primeval fireball is born, expending the energy derived from the collapse of the old universe on the expansion of the new.

As the universe is a closed system which never loses energy to its surroundings, this cycle of alternate expansion and contraction is eternal. It had no beginning and will never end. At first this may seem hard to accept, but the more you consider it the more you will find that it is unavoidable, as any possible alternative will also involve some eternal entity. Solutions proposed to eliminate this problem (usually by invoking magic!) do not solve it at all, but naively circumvent it by brushing it under the carpet.

There are people who say that as the number of cycles and therefore the number of universes created is infinite (though they are merely the same universe infinitely recycled), all sorts of strange worlds will eventually occur. For example, a world where there was no sudden extinction of the dinosaurs at the end of the Cretaceous, or the Normans lost the Battle of Hastings. Although there may be some degree of truth in this (after all, if something is possible then it is only a matter of time before it becomes inevitable) we are now trespassing into the realms of science fiction and it is futile to pursue such speculations further.

"The changing of Bodies into Light, and Light into Bodies, is very conformable to the Course of Nature, which seems delighted with Transmutations."

—Isaac Newton

II

The Spark of Life

It would be very difficult to convince a tribesman from the central highlands of New Guinea that the 'iron birds' which pass overhead from time to time do not work by magic or the agency of demons and spirits (even aircraft mechanics talk jokingly of 'gremlins' when trying to trace some elusive mechanical fault). An explanation based solely on mechanical principles would seem incomplete, lacking in some vital ingredient which would confer the spark of life.

The engineer on the other hand sees these things in terms of cylinders, pistons, crankshafts, sparking plugs and aerofoils, and this provides him with a complete explanation. Any other explanation would be superfluous. The engineer has no use for magic—he needs it like he needs a leaking radiator, a broken propellor blade or a burned-out bearing.

So it is with the mechanism of life itself. Though it may seem like magic to the layman, the biologist sees it as a collection of atoms and molecules working together in synchronised harmony like the cogs and pistons of a machine (though infinitely more complex than an aeroplane). Marvellous and wonderful, even awe inspiring life may be, but has no more need of magic than any man-made machine and supernatural explanations are a hindrance rather than an aid to understanding.

In the beginning was a great cloud of interstellar gas and dust hundreds of light years in diameter, not unlike many similar clouds to be seen in our galaxy to this very day. The great nebula in Orion is a typical and well known example. Though composed mainly of the primordial hydrogen and helium left over from the Big Bang, this cloud or nebula was enriched by considerable quantities of heavier elements spewed into the inter-stellar medium by generations of exploding stars (novae and supernovae) where these elements had been created, built up step by step from the original hydrogen. Turbulence within this cloud, probably produced by a nearby supernova exploding with the violence of countless hydrogen bombs, caused swirls and eddies where local concentrations of matter became sufficient to bring gravitational forces into play.

Gravitation is the weakest of the four fundamental forces possessed by matter at a sub-atomic level. The strongest of these is the strong nuclear interaction which binds together the particles of the atomic nucleus and is trillions of times stronger than gravitation, but its range is so short that it is effective only within the nucleus itself.

The second most powerful of these fundamental forces is the electromagnetic force which, though reaching to infinity, is largely cancelled by the fact that there are on average equal numbers of both positively and negatively charged particles in the universe, though there may sometimes be small local anomalies. The electromagnetic force is responsible for light, magnetism, chemical reactions and the holding together of atomic and molecular structures. It also gives the effect of solidity to the otherwise tenuous substance of the world we see around us.

Next comes the weak nuclear interaction, still many billions of times stronger than gravitation. Even shorter in range than the strong interaction, the weak interaction is only effective within the nucleus itself and therefore not apparent outside the world of the atom. It is responsible for beta decay and some particle collision phenomena.

Finally we have gravity itself, the weakest of the four forces which, though infinite in range (diminishing like light with increasing distance according to the inverse square law), is infinitesimal in its effects between individual atoms. Only in very large concentrations of matter do the cumulative effects become noticeable and in very large masses at high densities these effects become overwhelming, giving rise to such strange and spectacular phenomena as pulsars and black holes.

Once the turbulence in the primeval nebula had produced sufficient local concentration of matter, therefore, an irreversible gravitational bandwaggon effect was thereby set in motion, gathering force as the nebula contracted. The eddy, once beyond a certain density, could not disperse because of the increased effectiveness of gravitational attraction between its constituent particles.

The dense swirl of gas and dust, spinning ever more rapidly as it contracted, formed a proto-star, a dark globule of relatively high density nebular material perhaps only a couple of light years across. Dark globules of just this kind have been identified and photographed by astronomers, and are believed to be stars in the making.

At this stage collisions between particles of the collapsing cloud became frequent, especially at the centre, causing the proto-sun to heat up and begin radiating in the infra-red. This ball of gas and dust was still quite rarefied in the outer regions, and contained minute local irregularities in the form of condensates from the jets and filaments injected into the primeval nebula by exploding stars. These accreted frozen gas and dust to become comet-like bodies, agglomerations of meteoritic iron, nickel, silicon, carbon and small quantities of other elements in the form of dust and boulders embedded in a matrix of ice and frozen gases only a few kilometres in diameter.

The dark proto-star, still cool and tenuous except at the centre, became flattened as it gradually assumed a lenticular or cartwheel shape with a denser hub or nucleus. This was a result of more rapid rotation as the cloud contracted. The phenomenon is well seen in skaters, who bring their arms close in to the body in order to make themselves spin faster, and is due to conservation of angular momentum. The dense centre of the 'cartwheel' was the fore-runner of the sun as we now know it. The thin disc of material in orbit around it can be compared with the skater's flying skirt, and was eventually to become the planets, asteroids and comets.

Gravitational contraction and pressure in the central hub or core, aided by collision with infalling debris swept up from the nebula, soon brought about a rise in temperature sufficient to cause radiation of visible light or incandescense. At first this radiation was still partly hidden by the swirling fog of dust and nebular gases which shrouded the infant sun, but this was gradually swept clear by planets and comets and blasted away by the solar wind. The solar wind consists of fast protons (200–300 miles per second) streaming continuously from the sun in all directions, and is responsible for magnetic storms, radio interference, auroral displays and other phenomena here on Earth. Acting upon a large sail, much in the manner of an ordinary wind, it is still strong enough to be used as a means of space travel, but in the early days of the solar system the solar wind is thought to have been considerably stronger than it is now.

Most of the hydrogen in the original nebular eddy had been swept up by the dense central core. Within a few thousand years temperatures at the centre of the glowing proto-star became high enough to initiate thermonuclear reactions. This is essentially the same process which powers the hydrogen bomb, but because it takes place only at the centres of stars, and is self regulating for reasons which we will not go into here, the stars do not usually explode. Under the influence of terrific heat, hydrogen nuclei collide and fuse. When four have fused together they become an atom of helium with a total mass about 0·7 percent less than the four constituent particles. This vanishing mass is released as energy according to the well known equation $E=mc^2$, thus lighting a beacon which has shone steadily since for five billion years and continues to produce radiant energy at the astonishing rate of four million tons per second!

Meanwhile the dense mass of cometary bodies orbiting the star was undergoing repeated collisions because of their increased concentration and velocities, thus aggregating into larger bodies—the planets and planetesimals. Initially the planets were rather like miniature suns, having a large gaseous component which included hydrogen, and becoming marginally incandescent because of gravitational contraction and incessant bombardment by meteorites, planetesimals and comet-like bodies. Energy released by the decay of radioactive elements was a further factor in promoting high temperatures, but the essential difference was that owing to their small size heating of the planets was insufficient to initiate the fusion reactions which power the sun. On Earth, most of the heavy metals (largely iron and nickel) sank to the core during the semi-molten stage, leaving the lighter rocky materials to form the crust and mantle.

Because of their small masses, internal heating and proximity to the sun, the volatile hydrogen and other light atmospheric constituents of the innermost planets were soon boiled away by solar radiation unless bound to heavier elements in the form of water and other compounds. The giant outer planets (Jupiter, Saturn, Uranus, and Neptune) have largely retained these light elements, however, because of their powerful gravitational fields and greater distances from the sun. Jupiter, especially, is regarded by astronomers as a preserved sample of the primeval nebula.

At this early stage in the history of the solar system the planets were subjected to a ferocious bombardment by meteorites, comets, and planetesimals, some of which were of considerable size. In this way, by a process of accretion, the solid cores of the planets were built up from relatively modest beginnings. Traces of the final stages of the bombardment are clearly visible to this day on the surfaces of the moon and Mercury, and to a lesser extent on Mars and the satellites of Jupiter and Saturn (some if not all of which are themselves captured planetesimals). Collisions with such bodies became fewer as the inner reaches of the solar system were swept clear.

The remnants of the primeval swarm of comets are now largely confined to the outer reaches of the solar system, as might be expected, orbiting the sun at distances of up to quarter of a light year and sometimes taking millions of years to complete a single orbit. A few of the innermost members of this outer ring (Oort's Cloud as it is known, after the Dutch astronomer who first postulated its existence) are sometimes dislodged from their stable orbits by the giant planets, causing them to hurtle inwards towards the centre in highly elliptical sun-grazing orbits.

As they approach the sun their frozen gases are boiled off by solar radiations, fanning out into the long streamers which we see as comets. The solar wind, a stream of particles radiating from the sun, blows these streamers outwards, which is why cometary tails always point away from the sun regardless of which direction the comet is travelling. Swinging around the sun at high speed they then shoot off into the darkness of the outer regions, perhaps to return again in tens, hundreds, or thousands of years' time.

In 1983 something closely resembling an Oort cloud was detected in orbit around the bright star Vega. Such bright stars burn themselves out in a relatively short period of time, so there is no prospect of advanced life forms arising in a Vegan planetary system.

The asteroid belt, a ring of small moon-like objects and debris which circles the sun between the orbits of Mars and Jupiter, is the remains of a planet which either failed to form or was disrupted by impact in the early stages of formation.

With the final phase of the meteorite bombardment we have now reached a significant stage in the history of the Earth, so let us examine our planet as it existed nearly five billion years ago. Having cooled slightly from its recent marginal incandescence a thin crust had formed, pierced in many places by volcanoes. Most of the lighter gases had been driven off into space by the heat, but the planet still held a substantial atmosphere augmented by outgassing from volcanoes and the occasional arrival of comets bearing their burdens of frozen gases. A bleak lunar or Martian-like landscape extended from one horizon to the other, for although the interplanetary medium had largely been swept clear of debris the most recent craters were fresh and had yet to be erased by weathering. Meteorites were still of frequent occurrence by today's standards, but far less so than formerly.

Soon after this solid crust had formed and the atmosphere sufficiently cooled, torrential rains began to lash the cratered surface, often dissolving immediately into hissing clouds of steam which rose again into the sky. But in the cooler places, perhaps around the poles, great warm lakes began to fill the craters, becoming deeper and wider as countless centuries of rain began to form the oceans which were eventually to cover three quarters of the globe. Finally, about four and a half billion years ago, lifeless seas of dilute soda water covered many millions of square miles.

The stage was now set for the emergence of life, but in a very different world from the one we now know. For a start, the infant Earth spun more rapidly on its axis, so making a shorter day. Daylength was probably around 15 hours and there were more than 500 days in the year. The moon, which now turns one face permanently towards the Earth, then rotated slowly on its axis so that both sides could have been seen from Earth had there been anyone here to see them. It would also have seemed slightly larger, as it was then closer to the Earth. The same tidal energy which has since slowed the rotation of the Earth was captured by the moon and has boosted it into a higher orbit. Some of the most prominent lunar craters had not yet been made, and a thin halo of atmosphere probably lingered over its rocky surface. Even the stellar constellations would have been quite different from those of modern times and the night sky would have been almost unrecognisable.

On Earth the atmosphere was many pounds per square inch denser than it is today (perhaps as much as 3 atmospheres), several degrees warmer, and composed of gases poisonous to most animals now living. Of our modern life-forms only anaerobic microbes and perhaps a few plants could have tolerated this early atmosphere. In order to deduce what these primeval gases were it is useful to examine the atmospheres of some of our sister planets, which are comparatively little altered from their primeval state.

Jupiter, the giant planet beyond the orbit of Mars, still has much hydrogen and helium in its atmosphere, which is the least evolved of all planetary atmospheres and can be taken as a fair approximation of the primeval nebula. It is therefore representative of the primary atmospheres of all the planets, but these light gases have long since vanished from planets closer to the sun. In addition, the atmospheres of both Jupiter and Saturn contain significant amounts of two heavier, less volatile compounds of hydrogen: ammonia (NH_3), a compound of hydrogen and nitrogen, and methane (CH_4), in which hydrogen is combined with carbon.

The atmospheres of both our neighbouring planets, Venus toward the sun and Mars in an outer orbit further away from the sun than we are, consist of more than ninety percent carbon dioxide (CO_2), and this provides us with a further clue. When the primary (pre-bombardment) atmospheres of the inner planets were dissipated by the effects of heat and the solar wind only heavy gases such as CO_2 would have remained.

To this was added the products of volcanic outgassing—mostly carbon dioxide and water vapour, with traces of nitrogen and a few other gases—and the volatiles (mostly methane, ammonia and water) brought in by comets in the final stages of the accretion process.

This produced a secondary (post-bombardment) atmosphere which would have been similar on all three terrestrial planets (Mars, Earth, and Venus), containing large quantities of carbon dioxide, methane, ammonia and water vapour. In the upper atmosphere ultra-violet light acts upon methane (CH_4), ammonia (NH_3) and water vapour (H_2O) in a process known as photo-dissociation, which disrupts the molecules. The light hydrogen disperses into space, while the freed oxygen reacts with carbon to form a net residue of CO_2 and nitrogen. This is exactly what we find today on Mars and Venus, which may thus be said to possess evolved secondary atmospheres.

Bearing in mind the common origin of all the planets, Earth too must once have had significant quantities of these compound gases which although lost or broken down in comparatively early times must still have been important constituents of the Earth's primeval atmosphere. Evidence of the high carbon dioxide content of Earths primitive atmosphere is to be found in the huge deposits of fossil fuels and limestone rock (calcium carbonate). The billions of tons of carbon contained within these deposits once existed as a gas in the form of atmospheric CO_2.

Gradually, in the process which reached a peak about 350 million years ago during what is known as the Carboniferous period, this enormous quantity of carbon was extracted from the atmosphere by oil and limestone-forming marine organisms, and by the great Carboniferous coal forests which, after releasing the oxygen content of the CO_2 back into the atmosphere (hence today's abundance of free oxygen), buried the carbon in the coal seams. If all the oil and coal in the world was burned and all the carbon from the limestones released into the atmosphere we would revert to an atmosphere approaching that of three and a half billion years ago, of a sort still found on Mars and Venus, made up largely of CO_2. This will never actually happen, of course, because only a small percentage of oil and coal is recoverable and there is no equivalent process for extracting carbon from limestone.

Interesting evidence has recently come to light from the Voyager I examination of Titan, the giant satellite of Saturn which is larger than the planet Mercury. The gas-releasing effect of Titan's small gravitational field (about one fifth that of Earth) is almost exactly cancelled by the gas-retaining effect of its greater distance from the sun, and seems to have given Titan an atmosphere resembling that of Earth at about the time that life began. This atmosphere has been preserved in a state of deep freeze, so the similarities do not extend to temperature. Preliminary measurements indicate a ground pressure of between one and two atmospheres and the presence of hydrogen, methane, and free nitrogen. Traces of interesting and perhaps significant organic molecules have also been observed, but so far no sign of CO_2 or ammonia. This is not altogether surprising, as both these gases freeze solid at temperatures well above that of the upper atmosphere of Titan. The surface temperature is not known, but is likely to be cold enough to have frozen carbon dioxide and ammonia out of the atmosphere completely.

We may never be sure of the precise combination of gases involved in terms of exact percentages, but there are several indications that the primeval atmosphere which cocooned the lifeless cratered rock and soda-water seas of Earth four and a half thousand million years ago was up to 50 percent CO_2 (hence the soda-water seas!) The remaining fraction was mostly nitrogen, ammonia, and methane, with perhaps a dash of hydrogen and helium. Amounts of the inert gases argon and neon would have been insignificant, as is the case today. Argon in particular has only built up to its present one percent after billions of years' radioactive decay of potassium 40 in the Earth's crust. One of the most important differences in this early atmosphere was the lack of free oxygen, which would have been well under one percent of the total. This would have affected the amount of ozone (a tri-atomic molecular form of oxygen) forming the ozone layer which now filters out much of the sun's ultra-violet radiation in the upper atmosphere, and so allowed a much greater amount of these energetic wave-lengths to reach the ground.

Despite the fact that the infant sun was significantly cooler than it is now the 'greenhouse effect' of large quantities of carbon dioxide and water vapour, together with the thinner and warmer crust of the Earth, gave average temperatures several degrees higher than we have today, with the result that the polar caps were smaller and perhaps at times non-existent.

By the end of the violence and upheaval which characterised the formation of the Earth a sufficiently stable and benign environment to become the cradle of life had formed by about four and a half billion years ago, although the planet was as yet lifeless and still subject to bombardment by comets and meteorites. Although most of the Earth's surface would have been ocean-covered as it is today, the distribution of land and sea would have been very different. None of today's continents would have existed and dry land may have formed a single superconti-nent, a sort of proto-Pangaea, with here and there volcanic islands rearing up out of the sea. Large inland crater lakes would sometimes have been the size of small seas, and owing to their higher rate of evaporation would have had a higher mineral content than the ocean proper, especially in volcanic regions where particularly rich chemical soups would have formed.

On the shores of cratered deserts great rivers of lava ran hissing into a blue but lifeless sea, sending up gigantic clouds of steam. The dilute soda-water of this ancient sea would have been almost fresh and quite drinkable, especially as there were no microbes to contaminate it. There would have been very little salt, which has only accumulated in today's oceans after billions of years of leaching from the land.

The conditions required before life could appear involved (a) a plentiful supply of water at temperatures somewhere midway between

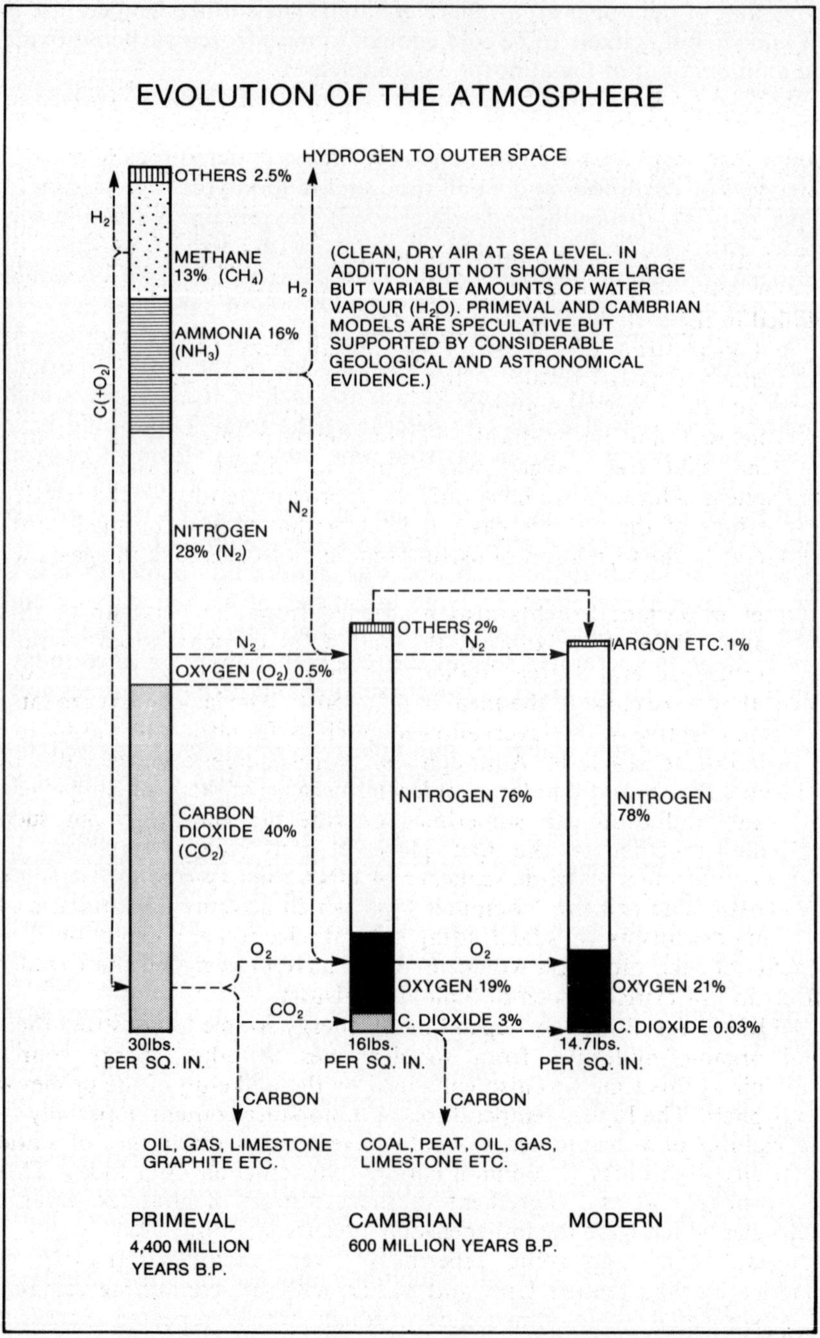
EVOLUTION OF THE ATMOSPHERE
HYDROGEN TO OUTER SPACE
OTHERS 2.5%
H_2
METHANE 13% (CH_4)
H_2
(CLEAN, DRY AIR AT SEA LEVEL. IN ADDITION BUT NOT SHOWN ARE LARGE BUT VARIABLE AMOUNTS OF WATER VAPOUR (H_2O). PRIMEVAL AND CAMBRIAN MODELS ARE SPECULATIVE BUT SUPPORTED BY CONSIDERABLE GEOLOGICAL AND ASTRONOMICAL EVIDENCE.)
AMMONIA 16% (NH_3)
$C(+O_2)$
N_2
NITROGEN 28% (N_2)
OTHERS 2%
N_2
N_2
ARGON ETC. 1%
OXYGEN (O_2) 0.5%
NITROGEN 76%
NITROGEN 78%
CARBON DIOXIDE 40% (CO_2)
O_2
O_2
OXYGEN 19%
OXYGEN 21%
CO_2
C. DIOXIDE 3%
C. DIOXIDE 0.03%
30lbs. PER SQ. IN.
16lbs. PER SQ. IN.
14.7lbs. PER SQ. IN.
CARBON
CARBON
OIL, GAS, LIMESTONE GRAPHITE ETC.
COAL, PEAT, OIL, GAS, LIMESTONE ETC.
PRIMEVAL
4,400 MILLION YEARS B.P.
CAMBRIAN
600 MILLION YEARS B.P.
MODERN

boiling and freezing point (100° and 0° Centigrade respectively). (b) An atmosphere rich in carbon, hydrogen, and nitrogen, plus adequate mineral deposits of phosphorous and trace elements. The presence of carbon was particularly vital, as it is the only element with the requisite chemical properties to build up the complex chains or polymers so characteristic of living things. The abilities of silicon, the next most versatile element, are far inferior. Another important factor in this early atmosphere was the lack of free oxygen. Oxygen is poisonous to very primitive life forms, which do not depend on it for energy. (c) The presence of suitable catalysts to promote the necessary chemical reactions and (d) abundant energy sources for the synthesis of large molecules from smaller ones. There is evidence that all these conditions were amply fulfilled in the early history of the Earth.

Geological strata of the early Precambrian formed from water-borne sedimentary deposits testify to the presence of water, and sedimentary rocks of this age containing elements which readily oxidise are nevertheless found un-oxidised or incompletely oxidised. This supports the view that free oxygen was scarce or absent in the primeval atmosphere. Although we have only an approximate idea of the relative proportions of the various gases it is clear that they were much more conducive to the formation of useful organic molecules than the gases we have now.

Other important benefits also proceeded from the make-up of this early atmosphere. For example, the high CO_2 content raised average temperatures several degrees higher than they are today so that in most places they were close to the ideal of 40°—50°C. The lack of oxygen (and hence an effective ozone layer) allowed much useful ultra-violet radiation to penetrate to sea level. Although the short, energetic wavelengths of ultra-violet are harmful to large molecules because of their ionising effect, this same radiation can sometimes provide the energy to put such molecules together in the first place. It also causes variations or mutations in the nucleotide sequence of DNA (de-oxy-ribonucleic acid), the genetic blue print or 'computer tape' which governs the structure of all living organisms, thus facilitating natural selection and evolution. The benefits of such radiation would therefore have outweighed the harmful effects in this critical epoch of biological history.

All life-forms ultimately depend on an energy source to construct their vital organic molecules from simpler ones. Another energy source available at this time was also enhanced by the make-up of the primeval atmosphere. The higher temperatures and moisture content, especially in the vicinity of volcanic vents, would have created discharges of static electricity (lightning), a common enough phenomenon even today. This was probably a vital ingredient in the synthesis of complex organic molecules which were the indispensable precursors of life itself.

Several years ago some experiments were performed (mainly in America by the chemists Urey and Miller) with jars containing versions of a simulated early atmosphere through which repeated electrical discharges were passed, representing lightning. Other energy sources used were heat and ultra-violet radiation. At the end of the experiments the contents of the jars were analysed and found to contain several important organic molecules, including amino acids, ribose, and the nucleotides adenine and guanine—some of the very building blocks of life!

Other possible energy sources include chemical energy (of a sort still used by a few primitive modern anaerobic bacteria) and geothermal energy. Apart from anything else heat permits chemical reactions to take place more briskly, especially where water is the solvent, which is why cellular metabolism comes to a halt at temperatures just below 0°C and resumes again (where the organisms are not killed) when the temperature is raised, reaching a maximum efficiency in most modern life forms at around 40°C. The bonds of many organic molecules are disrupted as temperatures approach boiling point, so that very few life forms, however primitive, can survive beyond 100°C.

The role of catalysts was probably a vital one, but presents no problem in our scenario as these materials are very common in the surface layers of the Earth's crust. A catalyst is a substance which facilitates chemical reactions without becoming directly involved itself, and may loosely be compared to a referee who, although almost indispensable in allowing play to proceed smoothly where it might otherwise grind to a halt, takes no active part in the game himself. Catalysts are widely used in the chemical and petroleum industries, and enzymes, which also have industrial applications, are their biological equivalents. Among the substances which may have had a role in building up the long chains of atoms (polymers) characteristic of living things are certain common clays which must often have lined the lakes and pools of the primeval Earth.

Put all these things together as they were in the early years of the lifeless Earth and you will soon have a rain of organic molecules collecting in the pools, crater lakes, and oceans. Molecules such as amino acids and nucleotides similar if not identical to the ones found in living creatures today would soon have created a rich organic broth—the legendary primeval soup.

Once created such molecules would be very long lived where they were shielded from ionising radiations, as there were by definition no living organisms to devour them as would be the case today. Complex organic molecules and polymers would therefore have become increasingly abundant in the 'primeval soup', and evaporation in shallow pools and crater lakes would have produced local concentrations of an even thicker broth.

It is important to realise that there was not just one primeval soup but many subtle variations of it which depended largely on local abundance of various minerals in pools of ground water. Even the atmospheric constituents were subject to local variation, as in the vicinity of volcanic vents or near the impact of large comets and their cargoes of volatiles. The Tunguska explosion which levelled hundreds of square miles of forest in Siberia in 1908 is believed by most experts to have been the arrival of a small comet. Such impacts were commonplace soon after the formation of the Earth.

There is also evidence that both comets and meteorites carry quantities of useful and surprisingly sophisticated organic compounds, and may have delivered them to the Earth's surface. Which is not the same thing as saying that life arrived here from outer space. How much of the more complex organic compounds survived the fiery ordeal of atmospheric entry and impact, and whether they played an important part in early biological activity is debatable. But if this sort of chemistry is possible in the harsh conditions of space, how much more is it possible in the comparatively benign conditions which prevailed in the early history of our planet?

At all events, the multitude of chemical soups and temperatures which existed meant that Nature had an almost unlimited number of flasks in her laboratory and hundreds of millions of years in which to conduct her pre-biotic experiments. There is a law of chance which states that where a given event is possible it is only a matter of time before it becomes inevitable, and if there was one commodity which abounded in the world of the early Precambrian it was time. In the words of the particle physicists: "What is not forbidden (impossible) is compulsory." Which has grave implications for the eventual use of our stockpiles of nuclear weapons!

The final synthesis of a self-replicating macro-molecule probably depended on a very unlikely chance combination of half-a-dozen sorts of smaller molecules in a long repeating sequence or chain. But what is unlikely in a laboratory flask within a time-span of days becomes quite probable, even inevitable, among myriads of subtly varying pools and lakes over periods of millions of years.

It is a pretty safe guess that the structure of this self-replicating molecule was not far removed from present-day DNA or RNA, its direct descendants, and modern amino acids and nucleotides. Therefore the key elements involved were carbon, available from methane and carbon dioxide (CH_4 and CO_2), nitrogen, available from free nitrogen and ammonia (N_2 and NH_3), plus hydrogen and oxygen. It is important to emphasise that there is nothing magical or mysterious about any of these elements, which are among the commonest and most abundant in the universe. We have already seen how they were put together in the interiors of stars from simple protons which, together with their attendant electrons, have been and sometimes still are created from pure energy.

Hydrogen and oxygen, though very scarce as free elements, were abundant in the form of compounds, especially as oxide of hydrogen or water (H_2O). Phosphorous would also have been necessary if the molecular side-chains for the nucleotide ladder were similar in form to today's, and this phosphorous would have been present as phosphates dissolved in pools of ground water.

Eventually, somewhere in the first 500 million years of the early Precambrian, an event occurred which was probably the most historic in the entire history of the Earth—the first self-replicating molecule, our direct ancestor, was synthesised in the primeval soup. Life had begun!

By dating methods involving rates of radioactive decay (remember the characteristic half-lives of unstable elements which we discussed in the previous chapter?) we know that the formation of the Earth by accretion of interplanetary debris was completed about four and a half billion years ago. There is also firm geological evidence of relatively advanced microscopic forms of life (though still only single celled) from strata laid down at least three and a half billion years ago (3·8 billion years has been quoted for some of them). As these earliest fossil life-forms, though tiny, were already well advanced on the road of evolution, they must therefore have evolved from still earlier and more primitive ancestors. There is no escaping the conclusion that somewhere between the formation of a suitable benign environment and the advent of the primitive virus-like organisms which were immediately ancestral to the earliest micro-fossils, life came into being upon Earth; not by magic but by well understood physical processes.

Bearing in mind the tediously slow progress of evolution in the early Precambrian and the comparatively swift decomposition (on a geological time-scale) of a favourable atmosphere by the mechanism of photo-dissociation, it seems quite likely that the emergence of life came about with what most astronomers and geologists and perhaps most biologists would see as extraordinary rapidity. Whilst the exact details are open to speculation and may never be known with absolute precision, this line of reasoning suggests that the great event may already have occurred as far back as 4·2 billion years B.P.(before present), within a mere 200 million years or so of the onset of favourable conditions.

Let us imagine that we can travel back in time to visit the historic site where life began on that momentous day. (I should perhaps mention that in reality time-travel *backwards* into the past is impossible, mainly because of the relationship between cause and effect. It is also forbidden by relativity, which is very fortunate. The mind boggles at the consequences if we should inadvertently snuff out some primitive ancestor before he had time to leave any descendants!)

Picture the scene on the shore of that great soda-water sea. A great desert, not unlike those to be found on Mars to the present day, stretches inland as far as the eye can see. The barren, rugged landscape, though lacking any sign of the greenery which will one day soften the harshness and beautify the Earth, has a desolate grandeur of its own which takes our breath away. It is easy to imagine how frightening and forbidding it would become during the hellish storms which are probably fiercer and more frequent than those we know from our own times. But today the weather is fine for a change as we prepare to alight from our time-machine. We must not forget to put on our breathing masks because of the poisonous atmosphere, otherwise we really will have our breath taken away! But they will be an irksome impediment in the prevailing high temperatures.

We feel a pain in our eardrums as the cabin is pressurised to equal the outside pressure of more than 2,000 millibars. This discomfort is soon relieved by the simple expedient of swallowing hard several times. The door opens to admit a blast of hot air, as from an oven, and we step out onto the sand. A nearby volcano belches dark, billowing clouds of gas and vapour which drift menacingly about the summit. A constant rain of complex organic molecules is synthesised by the electrical discharges which flicker incessantly within the cloud, and these are washed down by every passing shower into the sea itself and into the warm lakes of dormant craters. These crater lakes are rich in both minerals and organic molecules, and the mixture is gradually thickened by constant evaporation.

It is strange to reflect as we look around us that the whole landscape is as sterile as a surgeon's gown. Any soup or milk we boiled would remain drinkable for months just as surely as if it had been in a refrigerator, despite the sweltering heat. This is because there were no micro-organisms on the lifeless Earth until today, and it will be millions of years before they evolve into forms which can invade the land to cause decomposition and decay. Boiling would be difficult, however, because with less than one percent oxygen in the atmosphere combustion or fire as we know it is impossible and our camp stoves are unusable outside the time machine.

Even our lighters do not work. There is no chance of a smoke because of these damned breathing masks, but in any case our cigarettes would go out the moment they were lit. In view of the proportion of highly inflammable methane in the atmosphere this lack of oxygen is probably just as well. Thank goodness we will not be here for long. Perhaps we could boil our tea on a stream of red-hot lava from the volcano if we had to. There is a crater lake not very far away, so we stagger towards it, a sweaty task in our heavy breathing gear.

When we analyse the water we find several amino acids (the building blocks of proteins and enzymes) and at least two nucleotides, all molecules rich in nitrogen. Also in suspension are simple sugar (ribose) molecules synthesised from carbon dioxide, perhaps by the action of catalysts within the mineral rich pool. Among the minerals in solution are phosphates, which tend to clump together in short chains with a ribose and nucleotide molecule.

If we could place a droplet into some super electron microscope (which would need to be considerably better than any we have in modern times) we would soon have some interesting images on the screen, but very small and indistinct because of the high magnifications involved (if we have managed to get hold of a time machine then acquiring such a microscope should present no problems!). In order that we miss no vital detail our microscope also incorporates a device to slow down the picture, as events occur with phenomenal rapidity at atomic and molecular level. Among the throng of jostling particles the revered form of our very first ancestor drifts into view, and our highest magnification reveals a twisted ladder or staircase-like structure with chains of alternate sugar and phosphate molecules forming the uprights. Two nitrogen-rich nucleotides which we suspect are guanine and cytosine link together to form the steps or cross-pieces between each pair of sugar (ribose) molecules.

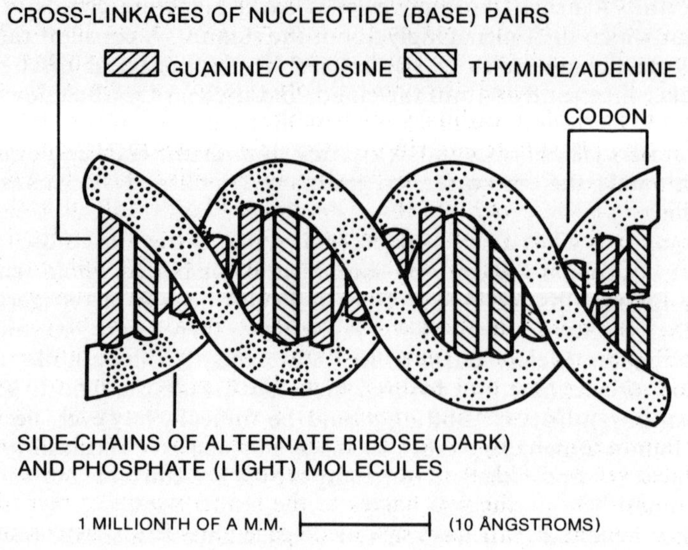

STRUCTURE OF DNA

When the ladder reaches a certain length it becomes unstable, and the nucleotide bonds begin to break at one end like an over-strained zip becoming slowly undone, dividing it through the centre of the rungs and splitting it into two mirror-image halves. These halves, continually jostled on all sides by water molecules in rapid thermal motion, are also subjected to bombardment by mineral and organic molecules. Those for which the half unzipped ladder-templates have an affinity are attracted from the jostling throng in the water by molecular forces, rather like pieces of steel drawn to a magnet, and tacked onto the newly broken edges.

The surprising thing is that they do not join up higgledy-piggledy but in a precise order which mirror-images the existing halves of the unzipped molecular ladder. A cytosine (?) molecule on the jagged edge of the template is buffeted by many kinds of small molecules, but having no affinity for them allows them to pass on by. Only when it collides with a passing guanine molecule do molecular forces come into play, and like a piece of steel passing a magnet the guanine molecule is plucked from the water and tacked onto the end of its counterpart, so completing a 'rung' of the ladder.

This automatic acceptance or rejection of spare parts according to their suitability is not so uncanny as it might first appear. The different nucleotides are not exactly equal but differ slightly in size, shape and electrostatic bonding so that one complements the other to make a perfect 'rung'. Any molecule of the wrong size, shape or electrostatic properties will not fit and is therefore rejected. The mechanism which moves the particles about is heat or thermal motion. When involving particles approaching pollen grains in size and therefore big enough to be clearly observed it is called Brownian motion and can be seen by anybody with the aid of an ordinary microscope.

In a similar way phosphate and ribose (sugar) molecules in the seething primeval soup attach themselves to passing nucleotides, each to their appointed place in such a way that they will eventually form the uprights or side-chains in the final duplication of the reconstituted molecular ladder. Soon the twisted nucleotide ladder has unzipped its full length, each jagged projecting nucleotide of the broken 'rungs' seizing its complementary partner from the teeming horde to repair new breaks as they appear. Replacement parts snap neatly into place with the same sort of precision that chlorine and sodium ions form precise and orderly ranks in a salt crystal or water molecules form a snow-flake. After continuous bombardment by exotic molecules during which all unsuitable candidates are rejected, it is not long before the transformation is complete. Lo and behold, the two halves of the ladder which so recently seemed broken beyond repair have salvaged spare parts from the passing current and built up replacement halves. We now have two 'ladders' where there was only one before!

In all probability this naked molecule would soon have been destroyed by adverse chemicals or radiations, in spite of the fact that there were as yet no predators. But this would not have mattered provided it had first replicated itself at least once. Before long a variation arose which was either able to synthesise its own protein covering from amino acids floating in the 'soup', or better still was able, like modern organisms, to build this armoured jacket for itself by synthesising its own amino acids. More likely this latter stage was only evolved at a much later date.

You may argue that such a molecule is not truly alive at all, but builds itself up from passing components in the manner of a snow-flake or a salt crystal. To some extent you are right, but the important thing is that here we have something, whether you call it alive or not, with the potential for infinite self-replication where conditions are right. It also has the capacity for infinite variation on a theme because of occasional mistakes or replicating errors (mutations), and therefore evolution and self improvement. The importance of size, shape, and electrostatic properties in governing molecular interactions is a feature of large organic molecules, especially proteins, as is the tendency to form long helical chains.

In the initially random sequence of the two nucleotides forming the 'rungs' we have the basis of a code, somewhat similar to the binary (two digit) code currently used in computers, which will carry advantageous information from one generation to the next, leading to ever increasing complexity and efficiency. The suggestion that there may have been only one base pair of two nucleotides involved is pure speculation, but based on the very strong probability that the first self replicating molecules

MOLECULAR STRUCTURE OF GUANINE AND CYTOSINE

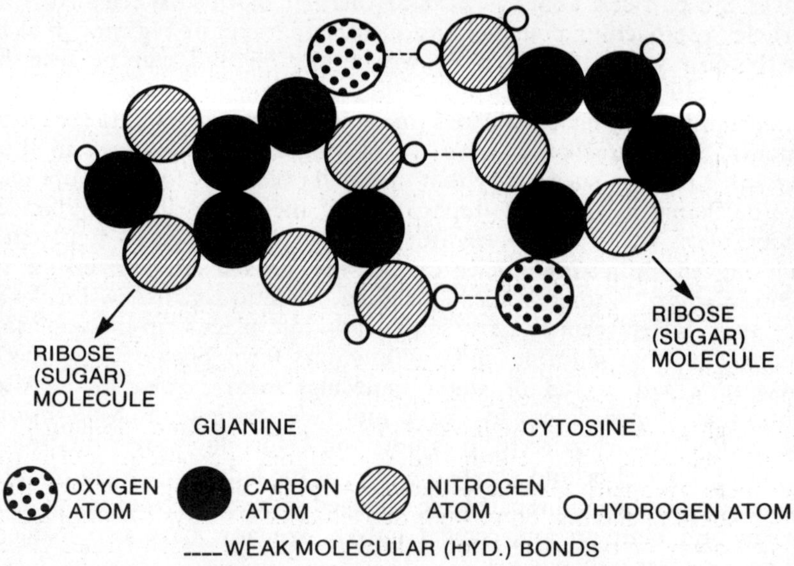

THE FOUR NUCLEOTIDES HAVE A NUMBER OF VARIANTS, INCLUDING THE CYTOSINE VARIANTS METHYLCYTOSINE AND HYDROXYMETHYLCYTOSINE. URACIL IS THE VARIANT OF THYMINE USED IN RNA. SEVERAL OTHER VARIANTS ARE KNOWN.

were of the simplest configuration possible. In modern organisms, which use *four* nucleotides forming *two* separate pairs (guanine-cytosine and adenine-thymine) this code has been broken and is well understood, though there remains a distinct possibility that it may have been slightly different in the earliest periods of biological history.

Each base or nucleotide forms a letter, and the code is spelt out in three-letter groups or 'words' which are usually known as codons. Because of the differences in electrostatic bonding, size and shape, members of one base pair cannot exchange partners with members of the other. Adenine must always pair with thymine (or vice versa), and guanine must always pair with cytosine, each bonded pair forming a link in the molecular chain.

Each three-letter group or codon codes for a specific amino acid—the building blocks of proteins and enzymes. For example, the sequence guanine-cytosine-cytosine, as read along the broken edge of a divided molecular 'ladder', spells out the instructions for the amino-acid alanine. Cytosine-guanine-guanine codes for the amino acid arganine, and cytosine-cytosine-guanine codes for proline. Other three letter sequences spell out 'full stops' or breaks in the message. However, the four nucleotides used in modern nucleic acids (RNA and DNA) have so many possible combinations that they are able to spell out replicating instructions for such complex organisms as a man or a bird—far more information carrying capacity than was necessary for the very first self-replicating molecules which were of the simplest possible construction.

These primitive virus-like organisms initially had a random nucleotide sequence, perhaps only a few dozen nucleotides long, subject to occasional replicating errors which produced variations on a theme. The causes of these errors or mutations were much the same as they are at the present time, namely radiations and chemicals which disrupted the sequence of the nucleic acid chain. Some of these variations were more stable than others or replicated more readily and were therefore advantageous, as they tended to predominate in the primeval soup at the expense of less fortunate combinations. We have here the beginnings of natural selection and evolution. Many mutations were unstable (and therefore harmful) and quickly died out, especially those which did not learn to fashion protective protein coverings from amino acids. But the rare advantageous mutations, when they occurred, passed on their advantageous information to their 'offspring' when they divided, and therefore perpetuated themselves and thrived.

Though governed by the laws of chance and the blind mechanism of natural selection which eliminated the failures, the random nucleotide sequences gradually became ordered, carrying meaningful information which could be handed on to their descendants. If relays of chimpanzees tapped away at typewriters for millions of years, most of their efforts would read as gibberish. But eventually, by pure chance or coincidence, one of them would type out a line from Shakespeare and if the relay could be kept going for long enough another would eventually type out a line and so on, until they had typed out the bard's entire works. Their human task masters, by destroying the gibberish, would thus simulate natural selection. This illustrates the way in which order can be created from randomness by blind chance and natural laws, a seeming paradox which often occurs in nature and produces an astounding effect of harmony and symmetry.

So it was with the nucleotide sequence of the first self-replicating molecule. It might have taken centuries of replicating errors shuffling the letters and millions of useless combinations before an advantageous message was spelled out by chance, but there was no shortage of time. The message might have spelled out how to pick up amino acids from the 'soup' and use them as weapons (enzymes) to attack and break down other self-replicating molecules and cannibalise their components, thus creating the world's first predator. Or how to synthesise proteins from amino acids and use them as a protective coating to resist chemicals, radiation or predators which might dismember the all-important nucleic acid. It is quite possible that somewhere along the line two specialised viroids might have joined forces in a primitive form of symbiosis, to create an organism more versatile and successful than either of its 'parents'.

Eventually, life forms arose which learned to modify existing nucleotides or gathered others from the 'soup', adding another nucleotide pair to their alphabet (assuming, of course, there were not two pairs from the outset), so greatly increasing the information-carrying capacity of the genetic code which was thus in more or less its modern form. This stage had almost certainly been reached by about 4 billion years B.P., shortly before the advent of the first unicellular fossils. (These micro-fossils were similar to today's bacteria and blue-green algae.)

It would have initially produced organisms similar to but simpler than today's simplest viruses, which are very simple indeed by the standards of most other life forms. It may well be that viruses are the modern descendants of such early pre-cellular life, just as today's most primitive unicellular organisms (bacteria and blue-green algae) are modern representatives of those found fossilised in three and a half billion-year-old rocks.

Viruses are extremely interesting, not only in their own right but also for the light they throw on the genetic machinery of living things and on the origin of life itself. Although considered by some biologists to be degenerate descendants of what were once cellular life forms similar to bacteria, there is good reason to believe that they represent modern forms of a pre-cellular organism only little more evolved than our very first ancestor. All modern viruses parasitise living cells, whether in human beings, other animals, bacteria or plants, but are probably very similar to free-living ancestors which thrived in the primeval soup. Indeed, they may well have affinities to the primeval macro-molecular predators previously mentioned.

When the primeval soup became over-exploited by competing life forms and inadequately replenished by nature, some organisms were forced to encapsulate themselves within a membrane or cell wall containing a version of the primeval soup which they were able to synthesise for themselves. Thus, when the 'soup' as such finally ceased to exist (which had probably happened by about 4 billion years B.P.) its passing was of little consequence. What could be more natural when other prey was becoming scarce than for the virus-like macro-molecular predators to attack this more advanced organism, within whose cell walls an encapsulated version of the primeval soup survived and still survives to this very day?

Viruses never got far beyond the stage of utilising pre-fabricated spare parts floating in the primeval soup, and are thus forced to parasitise its nearest modern equivalent in order to replicate. They have of course adapted to this new environment and learned to commandeer and make use of the cell's own reproductive machinery, which they do in a rather sinister way. Throwing off their protective protein jackets, they insert themselves into the cell's own nucleic acid (DNA or RNA) and 'order' the cell to produce copies of the virus instead of itself—in effect spoon feeding the virus with spare parts from the encapsulated fluid. Very soon an almost factory-like production line is under way, churning out copy after copy from the new blue-print it has been given. This often results in the death of the cell, which has virtually taken itself apart in order to provide material for the pirate virus production line.

A similar situation would arise if a robotic invader from outer space took over the automated and much publicised Fiat car factory in Turin. After the alien had substituted its own coded instructions for those in use by the factory's machines, automatons, and computers, the production line would be subverted to the production, not of motor cars, but of copies of the robot invader itself.

Another little trick which some viruses have perfected is to splice themselves into the cell's own DNA blue-print where they lie dormant, perhaps for years, disguised as part of the cell's own reproductive machinery, dividing when the rest of the DNA chain divides but showing no outward sign of their presence. There they remain, harmless, inactive, and unnoticed, until activated by some external stimulus such as ultra-violet light which unsplices them and lets them loose (the destabilising effect of ultra-violet light on nucleic acids has already been mentioned). They now come out into the open and proceed to take over the cell's reproductive and metabolic machinery in the orthodox manner. Herpes virus (*Herpes simplex*), which among other things causes cold sores, and chicken pox virus (*Herpes zoster*) which causes shingles, are examples of viruses using this strategy. Such viruses like to follow nerve sheaths, which seem to provide them with convenient highways. It may be that they are protected here from attack by the body's natural defences, the immune system, which is also unable to get at them when they are hiding among the cellular or chromosomal DNA.

The splicing and unsplicing of genes in the DNA nucleotide sequence is the basis for what is sometimes called genetic engineering, the ability to tamper with the cell's genetic blue-print, and viruses are among the agents used to bring this about. So you see, if life really was magic then genetic engineering, from which we will soon be reaping enormous benefits, would be impossible except, perhaps, by means of sorcery and witchcraft!

The simple and functional designs of viruses very often resemble man-made artifacts rather than living organisms, and this together with their mechanical style of reproduction and assembly has led some biologists to question whether they are alive at all. Viruses dis-assembled into their component parts (which individually are not alive) have automatically put themselves back together again when incubated in water, forming live virus. The process of self-assembly is similar to that described for the first primitive life forms in the primeval soup.

Several viruses have been crystallised in much the same way as salt, sugar and other inert materials can be crystallised, and in this form they can survive almost indefinitely like lifeless chemicals, but become infective and alive again when inserted into living tissue. Their astounding simplicity is emphasised by the fact that the genetic instructions for the very simplest viruses are only about 1,000 nucleotides long, and this contains only about 300 triplet nucleotide 'words' or codons. The codons dictate the structure and characteristics of living things by spelling out instructions for the production of proteins and enzymes from various combinations of the twenty or so naturally occurring amino acids.

As every schoolboy knows, the basic unit of heredity is the gene, which uses these proteins and enzymes to influence physical characteristics. The number of codons coding for a protein and therefore the number to the gene is variable, but averages about 90 for each protein or enzyme and about 150 to the gene. So the nucleotide sequences of the smallest viruses are sufficient to code for only about two average genes.

By comparison their primeval ancestors probably contained sequences of only a few dozen nucleotides, but the genetic information required in order to replicate a human being is *billions* of nucleotides in length. Some of these viruses are only 100 Angstroms in diameter (10 Angstroms = 1 millionth of a millimetre).

Perhaps some ultra-primitive virus may yet be found exhibiting primitive features which will throw further light on the structure of free-living ancestors, pointing the way to a more positive reconstruction of the very first life form—our primeval ancestor. It is quite possible that within the lifetimes of people now living this very molecule may be created in a laboratory flask—the most astounding conservation feat of all time: a living specimen of an ancestor which became extinct over 4 billion years ago!

One of the reasons very little work has been done on the problem of synthesising life in the laboratory is that there are unlikely to be any commercial benefits in doing so. The research represents the pursuit of knowledge for its own sake. But it will undoubtedly be done one day, and if the synthetic medium is a reasonable approximation of conditions prevailing in the primeval soup of four and a half billion years ago, there is no reason why the resulting virus-like molecule should not be an identical living copy of our first ancestor. The re-constitution of dis-assembled virus from lifeless components in a test-tube has already been done, but does not count as the creation of life because these components were of biological origin, as was the nucleotide sequence of the nucleic acid.

The role (if any) of Rickettsia in this scenario is not clear. Rickettsia are the organisms responsible for typhus and Rocky Mountain spotted fever. In size they are little larger than viruses, and seem to be transitional between viruses and bacteria. Whether or not they represent modern forms of a primeval proto-cellular organism (which bacteria undoubtedly do) is debatable, but I am inclined myself to think they do. There is also the possibility that they may be degenerate descendants of more advanced organisms. Despite the fact that they are among the most primitive living things, bacteria are relatively complex and far too advanced to have been the very first form of life. Somewhere along the evolutionary line there must have been more primitive ancestors which at least resembled Rickettsia, regardless of true relationships. Be that as it may, the science of macromolecular biology still has exciting prospects before it and will continue to throw light on the fundamental nature of life.

A curious phenomenon which has yet to be explained concerns the asymmetry or 'handedness' of amino acids, the building blocks of proteins which were utilised very early in biological history to provide protective coatings for macromolecular life forms. Amino acids in nature have two molecular forms which are at the same time identical but different. The apparent contradiction is explained when you look into a mirror and compare the image with the original. Or look down at your hands, and again you will see that they are identical but different—one left, the other right. So it is with the molecular structure of amino acids, the raw materials of proteins and enzymes which became increasingly important as life progressed through the viral stages and beyond.

Although left and right handed forms are about equal in amino acids of non-biological origin, it is a curious fact that in biological structures only the 'left handed' type are used. Though there have been attempts to attach great significance to this phenomenon, a probable explanation is that longer, more stable, and therefore more efficient chains (proteins and enzymes) could be built up using only one form (isomer) of amino acid instead of a mixture of both. The fact that left was 'chosen' in preference to right was probably due only to chance.

Incidentally, most viruses are host specific and cannot survive for long outside a living body. Bacteria are either host specific or habitat specific and few can reproduce themselves outside the environment for which they have evolved. This appears to rule out the eccentric theory that disease germs are brought to Earth by comets. I predict that no Earthly diseases (or for that matter, germs of any kind) will be found preserved in the Martian ice-caps, though they ought to be there if the theory is correct.

III

The Road to Darwin

The secondary atmosphere which had once rained complex organic molecules into the lakes and oceans was gradually modified, mainly by photo-dissociation and the depletion of the cometary swarm which had supplied the planets with volatiles, so that the primeval soup lost one exotic ingredient after another in successive stages until it finally ceased to exist, the whole process taking 'only' a few hundred million years. Ammonia and methane all but disappeared. It was this gradual impoverishment of the primeval soup which brought about the emergence of cellular organisms. The land areas of the globe still resembled a moonscape or the deserts of Mars, and the harsh conditions which discouraged exploration of new environments were only suitable for very primitive life-forms.

By the time the first fossil life forms appeared about three and a half billion years ago life had existed on Earth for at least half a billion years, but had still only reached the stage of bacteria and primitive unicellular plants known as blue-green algae (not to be confused with the much later and more advanced green or true algae). Collectively termed procaryotes, these are the simplest forms of cellular life, lacking the more complex metabolic machinery (organelles), and the well defined nucleus containing the DNA-bearing chromosomes in more advanced organisms. Despite this relative simplicity some of them, at least, are able to synthesise all the essential chemicals they need from non-biological sources in the environment to produce their own private reservoir of 'primeval soup' which they enclose within a membrane or cell wall. Life was to remain at a unicellular level for a further two billion years until the late Precambrian. Although outwardly the progress of evolution was very slow, significant strides were made in cellular structure, metabolism, and reproduction during this time.

The carbon dioxide content of the atmosphere moderated considerably during the early part of the Precambrian as the oceans became the planet's lungs where cyanobacteria and other primitive unicellular organisms broke down carbon dioxide and poured oxygen into the air. This was the beginning of what might be described as a tertiary or third stage atmosphere brought about by the action of living things, and therefore unique in the solar system. In this stage ozone began to increase in the upper atmosphere, cutting down the injurious amounts of ultra-violet radiation reaching the ground, and the stifling temperatures caused by the greenhouse effect were lowered slightly as the oceans consumed carbon dioxide.

The next major step in the progress of life was the evolution of the eucaryotic cell in the form of protozoa (amoeba-like animals) and simple unicellular plants, which seems to have taken place in the mid Precambrian about a billion years after the appearance of bacteria. This more advanced cell has the chromosomes (DNA) contained in a well defined central nucleus bounded by a membrane, and metabolic processes are carried out in discrete organelles as opposed to mere folds, pores and enzymes in the cell wall.

The obvious differences between plants and animals break down at a unicellular level and they exhibit a remarkable degree of similarity and overlap. Even the dividing line between unicellular and multicellular organisms is blurred, and some strange amoeba-like creatures called slime moulds, though mobile and capable of independent existence, are able to clump together to build a living pyramid which becomes recognisable as a multi-cellular fungus. This process parallels, perhaps pre-dates and may well have been the fore-runner of the cell migration which takes place in the embryonic stages of more advanced organisms to create their specialised form and functions and transforms caterpillar to butterfly in a chrysalis.

It has been widely speculated that the oxygen metabolising, energy consuming organelles known as mitochondria, found in both plant and animal cells, and the chlorophyll-containing chloroplasts which capture energy from sunlight in green plants, are a form of symbiotic relationship between procaryotic and eucaryotic cells dating from about this time. A well known and undisputed form of symbiosis is the amalgamation of algae and fungi which produces lichens. The theory goes that bacteria-like and algae-like procaryotes united with eucaryotic cells to their mutual advantage, and there they have remained to this day.

Particularly in the case of mitochondria there is some some circumstantial evidence to support this idea (for example, they contain their own DNA and the triple letter codons are very slightly different from those found in chromosomal DNA), but the case is far from proven. This sort of mystery tickles the fancy of reality-watchers far more than the question of whether there are fairies at the bottom of the garden, reincarnation, or whether pigs have wings, and similar artificial 'mysteries'.

Perhaps partly as a result of improving conditions, by the late Precambrian some one and a half billion years ago simple multicellular organisms began to appear. At first these were little more than colonies of near-identical cells, but gradually cells within these colonies began to differentiate, to adapt to specific tasks and form recognisable organs. This is the way in which the specialised organs of higher animals began.

About this time came the innovation which really enabled evolution to take off–the development of sexual reproduction. This mingling of genes from two separate individuals, a sort of genetic cocktail shaker, enabled a far greater variety of genetic combinations to be assembled, giving natural selection much more to choose from. In some cases it also has the effect of neutralising or suppressing potentially harmful genes. The normal method of reproduction among single-celled organisms is by

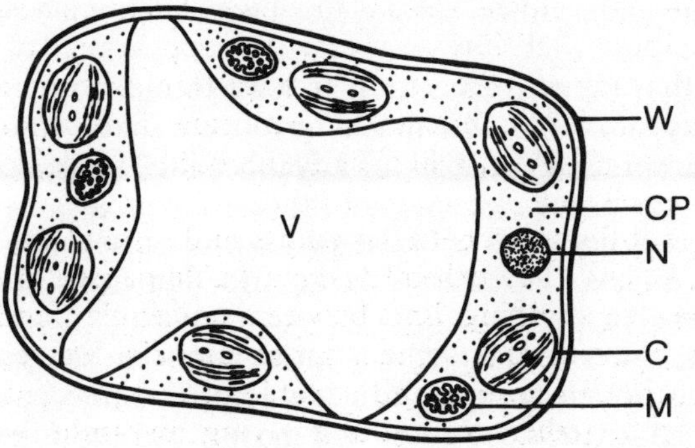

V VACUOLE (SAP)
W CELL WALL
CP CYTOPLASM
N NUCLEUS
C CHLOROPLAST
M MITOCHONDRION

TYPICAL PLANT CELL
SHOWING PRINCIPAL ORGANELLES

Bacteria-sized chloroplasts (C) and mitochondria (M) each have their own typical DNA. Are they descendants of ancient procaryotes in symbiotic association with eucaryotic cells?

means of binary fission or splitting in two—an extreme form of inbreeding. A few modern bacteria have developed a primitive and inefficient method of sexual reproduction but it seems unlikely that this was evolved by their primeval counterparts.

Well before the end of the Precambrian, jelly-fish, worms, and other simple multicellular animals began to appear. Shrouded and obscured in the mists of time this dimly seen period, a sort of biological Dark Ages, had been eight times longer than all the remaining geological periods put together. Throughout this time (and for many millions of years to come) the land areas of the globe had been barren, lifeless deserts. But evolution, though still slow, was gathering pace. It may be that the great surge and diversification of life which seems to have occurred at the end of the Precambrian was brought about mainly by the increasingly well oxygenated atmosphere and the advent of sexual reproduction.

The beginning of the Cambrian period ushered in a new era in the history of life on Earth. Creatures with large horny coverings forming easily preserved fossils began to increase and diversify. Hitherto such fossils had been very few, as soft bodied creatures and micro-organisms are preserved only in exceptionally favourable circumstances. Even so, another 100 million years were to pass before the emergence of the first vertebrates or backboned animals in the geological period known as the Ordovician, when in a line of descent from segmented worms, through forms resembling little soft bodied sea creatures called sea-squirts and the rather more fish-like lancelets, primitive jawless fishes evolved. The segmented character of modern fish, reptiles and mammals is not apparent until you examine the skeletons with their rows of repeated ribs and vertebrae.

It may be that the reason it took so long to make this important step was that conditions, though much improved from those of the early Precambrian, were not yet ideal for advanced life forms. It is just possible that by this stage primitive plants in the form of mosses had left the lakes and swamps and begun to colonise the damper places on land, but only later, in the Silurian, were they large and numerous enough to leave readily identifiable fossils.

The earliest fishes of the Ordovician were barely recognisable as such and resembled the modern hag-fish and lampreys, but in the succeeding Silurian around 450,000,000 years ago more advanced forms arose. Most of these were heavily armoured around the head and fore-parts, and some were of considerable size. The development of bony jaws was an important innovation and gave rise to the world's first recognisable carnivores (hence the popularity of armour plating). The greatest forests on Earth were still only a few inches high.

By the end of the Silurian life was still largely confined to the sea, though a few scorpion-like creatures and primitive flightless insects had started to invade the land, roaming the knee-high jungles of mosses and ferns. The earliest insects were shrimp-like creatures resembling the tiny 'silver-fish' sometimes found in dark cellars and closets.

The Devonian was chiefly an age of fishes when advanced forms resembling some modern species first appeared. Land plants developed enormously, producing the first trees and forests as we at present understand them, though these first primitive forests were made up of giant ferns and horse-tails which are now found in only a few tropical locations. The most important event which took place in the Devonian was the development of fishes which could leave the water and exist for short periods on land. These lobe-finned fishes, by a modification of the swim-bladder, were able to breathe air when on land, and combined several features found in modern lung-fishes, mud-skippers, and coelacanths.

Eventually they became so well adapted to living on land that they became almost independent of water, and the first amphibians had arrived. These resembled modern frogs, newts, and salamanders, especially the latter, and in the development of tadpoles, which initially resemble gill-breathing fish but ultimately adapt to existence on land, the evolutionary drama is re-enacted before our eyes. By the end of the Devonian the first of these newt-like amphibians had stepped onto dry land into a pleasant oxygen-rich atmosphere not very different from that which exists today.

This invasion of the land gathered pace in the Carboniferous period, when in lowland bogs giant horse-tails and ferns grew to the size of mighty trees, forming great swampy jungles and forests. The decaying trunks of these trees formed layers of peat which, buried beneath many feet of water-borne sediments, became compressed and carbonised, eventually forming coal. In this way the last excess of carbon dioxide was extracted from the atmosphere and separated into its component parts. The oxygen was returned to the atmosphere and the carbon buried in the coal seams. Coal formation was not restricted to the Carboniferous, however, though most of the coal measures date from about this time. Smaller quantities have been produced in all geological periods from the advent of the first forests in the Devonian right down to the present day. The formation of limestone, as well as petroleum and other hydrocarbons began in far earlier times and, notwithstanding periodic surges and declines, has continued ever since.

It should be pointed out that, contrary to popular opinion, the majority of photosynthesis and oxygen production has always taken place in the oceans and still does, not in the tropical rainforest. This is made possible by the teeming hordes of tiny single-celled plants which tint the oceans green. As nearly three quarters of the Earth's surface is covered by water their combined effects far outweigh those of the jungles and forests. These make a contribution of course, but are mainly valuable for quite different reasons.

Amphibians, with little opposition on land, went from strength to strength in the Carboniferous, developing a multitude of reptile-like forms larger and more impressive than any alive today. Their skins, however, like those of modern amphibians, were damp and easily lost water. Therefore they could not stray far from moist and shady places, which abounded in the Carboniferous coal forests. This was a major handicap, as it meant that they were virtually forbidden from exploiting the vast arid regions of the Earth's surface. They were also dependent on water for breeding, and in this respect still resembled fishes.

Insects too had developed into more or less modern form, and giant dragonflies, some with a three foot wing span, flitted between the giant horse-tails and palm-like tree-ferns of the damp coal forests. The aquatic larvae of dragonflies, mayflies and their relatives, or even the larvae of mosquitoes which are air-breathing for most of the time, suggest a shrimp-like ancestor for the insects.

The desert areas of the Earth were still virtually devoid of life, forming a vast ecological niche just waiting to be exploited. Nature abhors a vacuum. The next great evolutionary step was the emergence of the amniotic (solid shelled) egg which transformed amphibians into reptiles by releasing them from their dependence on water. The amniotic egg is in effect a small pond encapsulated by a shell in which the embryo can develop through its aquatic stages, nourished by a fatty food store or yolk. This, together with their more water-retaining skins, enabled the first reptiles to invade the dryer places where amphibians could not go. This made reptiles a very successful group indeed, which, despite competition from mammals, they continue to be right up to the present day.

By the late Carboniferous amphibians had developed advanced forms, somewhat crocodilian in appearance, which strongly resembled reptiles. Whether the first primitive reptiles had appeard by the end of the period is not absolutely certain, as the definitive features mentioned above are not easily preserved, but there is no doubt that early in the Permian true reptiles roamed the Earth.

Once established, the reptiles rapidly diversified and expanded. The days of amphibians as the dominant life form were numbered. This follows the usual pattern in nature where change accompanies exploitation of a new environment, and helps explain why some creatures evolve rapidly while others remain almost unchanged for many millions of years. Thus was inaugurated the Age of Reptiles, which lasted for 200 million years until the extinction of the dinosaurs, and during this period, which included the Triassic, Jurassic, and Cretaceous, reptiles of every imaginable size and form dominated the Earth. Not only on land but in the sea, and for a long while even in the air.

At around the time of the first reptiles plants too were becoming more like those we know today, as ferns evolved into cycads and cycads into primitive conifers. Cycads are palm-like trees which are still found in parts of the tropics.

A group of advanced reptiles known as mammal-like reptiles appeared surprisingly early in the Permian. The name is rather misleading, however, as these early forms were only like mammals in a few subtle skeletal features and not at all in general appearance, but were probably ancestral to true mammals. For much the same reasons as it is difficult to pin-point exactly the emergence of reptiles from advanced amphibians it is also difficult to define precisely when true mammals arose from advanced mammal-like reptiles. The first unequivocally mammalian fossils do not appear until the time of the first dinosaurs toward the end of the succeeding geological period, the Triassic.

These first mammals were insignificant shrew-like creatures which, probably because of competition with the predominant reptiles, they were to remain until the extinction of the dinosaurs over a hundred million years later. Being small they would easily have become torpid during cool nights, and so had evolved warm blood and an improved circulatory system in order to remain active. They were therefore probably nocturnal, as are most mammals today, and though hunted by reptiles during the day this gave them a significant advantage at night when reptiles become sluggish and inefficient. Modified scales provided insulation in the form of hairs. At this stage mammals probably still layed eggs, in this respect resembling modern monotremes (echidna and duck-billed platypus) which still retain several reptilian characteristics.

Turtles and crocodiles which, like dinosaurs, also originated in the Triassic, have changed very little up to the present day. Dinosaurs originated from an aquatic ancestor not unlike a crocodile, and in the succeeding Jurassic they evolved a multitude of forms which dominated the Earth. Not all of them were pea-brained dimwits and many were far more advanced than any reptile alive today. But as the great giants reached the height of their powers the little creatures who were eventually to succeed them peered out at them from the trees and undergrowth. The first of these were the primitive, mainly insectivorous mammals which have already been considered. The second was a small bipedal reptile closely related to dinosaurs (which were not all giants by any means).

The removal of the Earth's blanket of carbon-dioxide had paved the way for advanced life forms to appear, but it had also lowered the temperature of the Earth to something near what it is at present. With day length at about twenty-three and a half hours the temperature contrast between night and day was only slightly less than it is now and much more than in Precambrian times when, owing to the denser atmosphere and more rapid rotation of the Earth, it was negligible. As mammals had already discovered, reptiles become torpid and vulnerable at low temperatures and cannot remain active unless they stay warm. It has been speculated that in order to overcome this, several species of warm-blooded reptiles had evolved, including some of the dinosaurs. The early bird catches the worm, so in order to get off to a quick start in the morning one active group of small bipedal dinosaurs had evolved both warm blood and an insulating covering of specially modified scales. These modified scales we now call feathers, and their original purpose was not flight but insulation.

However, a genetic 'mistake' or mutation, an alteration to the DNA coded instructions brought about by background atomic radiation or chemical interferance, caused the fore-arm feathers of one individual to grow longer than was normal. Not only was this found attractive by the females, but came in very handy when fleeing from predators. The longer feathers not only gave better balance and aerodynamic stabilisation, but by beating the air when running also helped to propel him forwards. This useful trait was therefore passed on to his descendants, and some of the more well endowed of these found that with the long feathery tail acting as a parachute these aerodynamic stabilisers would permit exceptionally long jumps from one tree to another. This was also useful, not only for escaping from predators but for gathering food; therefore those individuals which inherited these features to an exaggerated degree survived and prospered to pass them on to future generations, whilst many of those with a lesser inheritance did not.

We have here an object lesson in evolution and the workings of natural selection—the survival of the fittest. My example is slightly over-simplified for the sake of clarity, but only a little less valid for that. A more scientific explanation would be that species are made up of populations of variable individuals whose success in the struggle for survival is reflected in the numbers of their offspring. Those whose genetic variation helps them adapt to their environment will flourish and leave numerous descendants, whilst those lacking these favourable characters will tend to be weeded out by natural selection—survival of the fittest—and leave fewer offspring. When biologists talk of 'the fittest' they are referring to *reproductive* fitness, the ability to leave larger than average numbers of descendants which grow to maturity. Therefore natural selection is usually a statistical phenomenon in which certain favourable characteristics which help a species to adapt become predominant in the population, and those which do not tend to die out.

Not surprisingly, it usually takes many generations before what started out as a statistical pattern of genetic inheritance becomes a standard feature of all individuals, so even the most trivial manifestation of natural selection can seldom be observed within the span of a human lifetime. The ultimate source of all variation is mutation, but the phenomenon of recombination—the crossing over of linked nucleotide sequences from one chromosome to another, whether of whole genes (intergenic) or parts of genes (intragenic)—means that mutations which have been accumulated but lain dormant for thousands or perhaps even millions of years can be activated. This is very fortunate, as in normal circumstances mutations are comparatively rare events.

Now to get back to our feathered reptile. From long jumps it was a simple step to short glides, and the development of the pectoral muscles and fusion of the clavicles to form a wish-bone soon permitted a laboured, flapping flight. The first birds, typified by Archaeopteryx, had arrived. The skeleton of Archaeopteryx is so reptilian, with its long jointed tail and toothed jaws, that at least one specimen was at first mistaken by palaeontologists for a reptile. The three other known specimens would probably also have been classed as reptiles had it not been for the fact that impressions of feathers had been preserved.

Many reptilian features are still evident in modern birds, and are visible proof of evolution in action. Although the teeth have vanished, the beak is a feature shared with several modern reptiles and many extinct species. The four toed foot structure with stout claws or talons is one shared with many dinosaurs, including the famous *Tyrannosaurus rex*. Like reptiles, birds lay eggs, have scaly feet and legs, and even the feathers are only modified scales.

The bipedal running posture is also shared by many reptiles, both living and extinct, and vestiges of reptilian fore-feet complete with talons are to be found in the skeleton of the wing. These grasping fore-limbs were well developed in Archaeopteryx, and one modern bird has retained them to a surprising degree. The young of the South American hoatzin are able to use these grasping 'hands' to climb about the branches of the nesting tree, just as Archaeopteryx did 140,000,000 years ago. Unlike Archaeopteryx, however, these fore-limbs lose their reptilian function in the adult bird. The long reptilian tail of Archaeopteryx has become shortened in modern birds to a vestigial structure known as the pygostyle. The use of gastroliths (stones swallowed to aid digestion) is another feature common to both reptiles and birds, and similarities extend even to microscopic cell structures.

Flowering plants, which include all the familiar modern broad-leaved trees, were probably beginning to emerge at about this time but do not reliably appear in the fossil record until much later, in the Cretaceous. Conifers were already well established by the Jurassic and most of the more ancient types of trees such as cycads and tree-ferns still quite common, so flowering plants may have been scarce and insignificant, growing in habitats where they did not easily fossilise. Reptiles were nearing the height of their development, and several were challenging

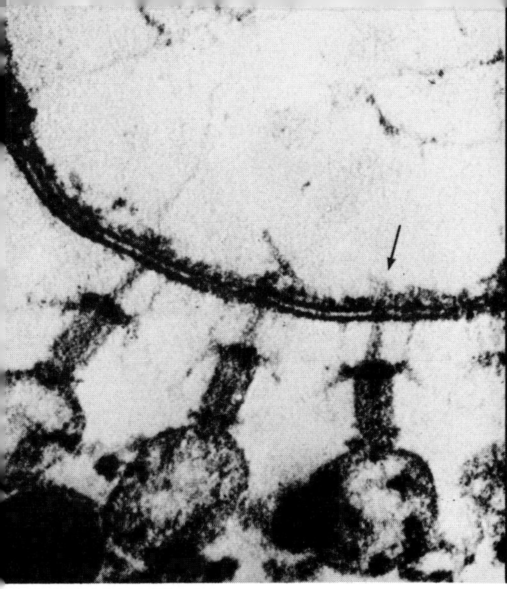

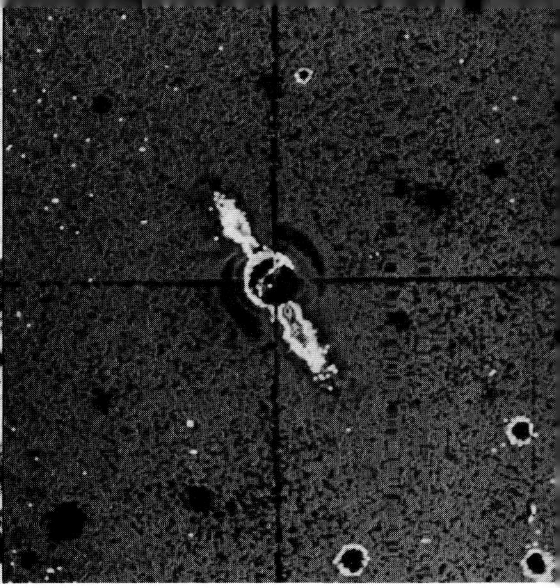

Above left: Mosquito-like bacteriophage virus injecting ribbon of DNA (arrowed) into host bacterium.

Above right: Charge-couple device (electronic) photograph of Oort cloud in orbit around the star Beta Pictoris. Several other examples are known, including the stars Vega and Fomalhaut. Believed to consist mainly of water ice, such clouds may well conceal planets.

Below: Electron micrograph of bacteriophage virus (×250,000), showing stacked disc structure of tail and the spiky base plate and mosquito-like 'legs' by which it anchors itself to host bacterium. The 'tail' is actually a syringe which contracts, driving a slender tube through cell wall of bacterium.

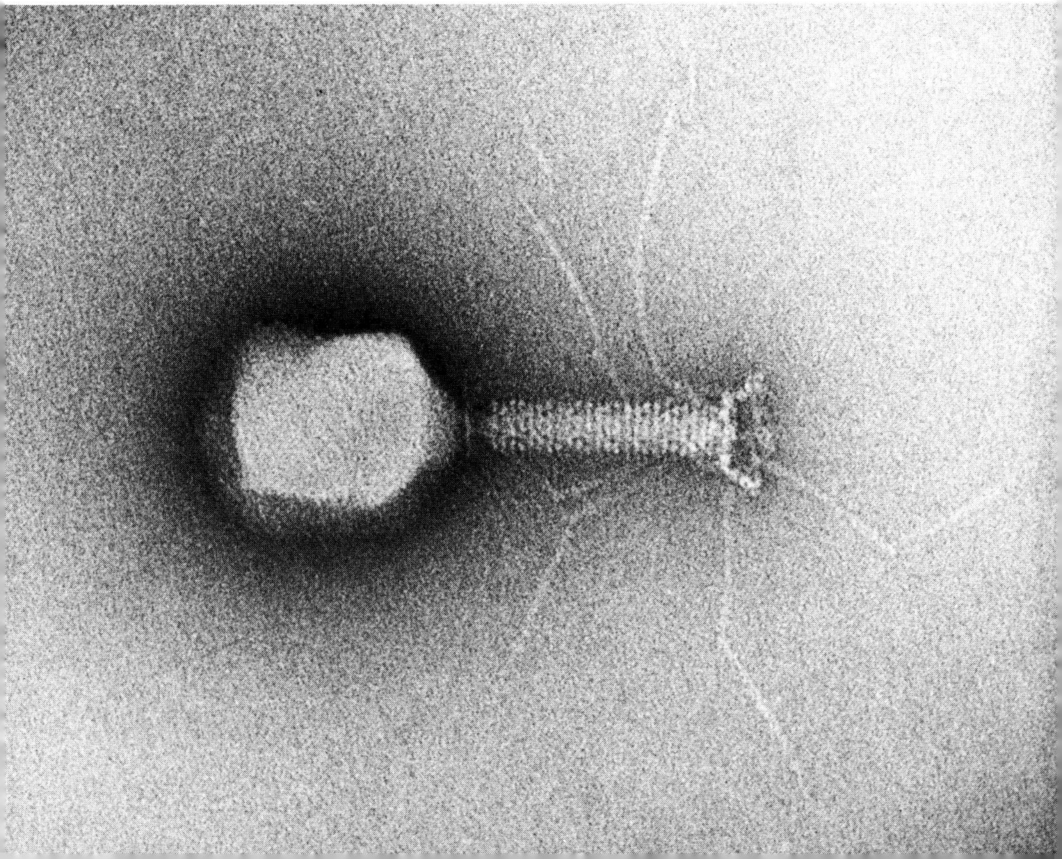

Australopithecus africanus, *the advanced hominid which was certainly very close to man and may actually have been man's immediate ancestor. Although the brain was that of an ape, the skeleton was very similar to our own.*

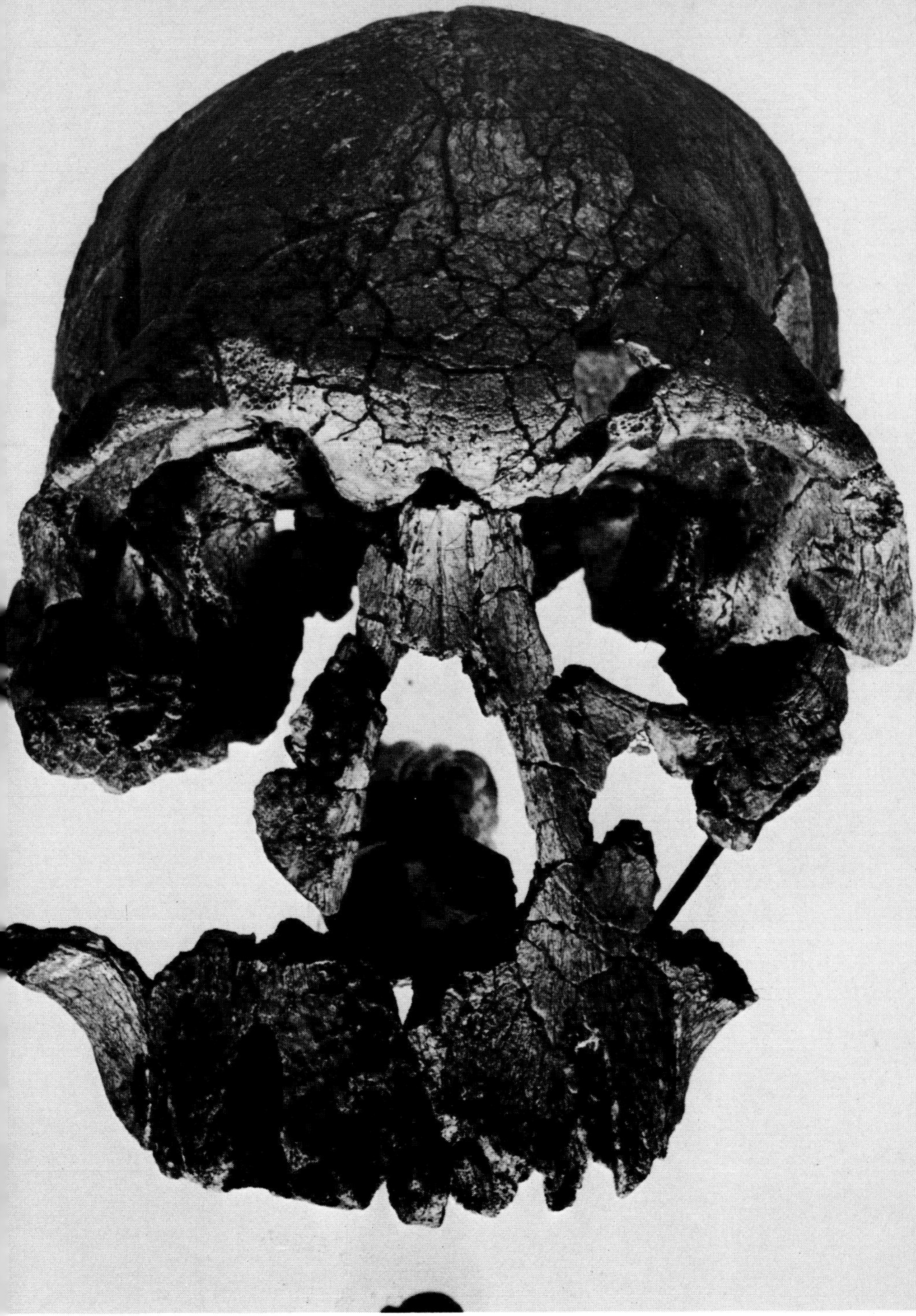

Homo habilis *(skull 1470 from Lake Turkana, Kenya), the earliest known species of man.*

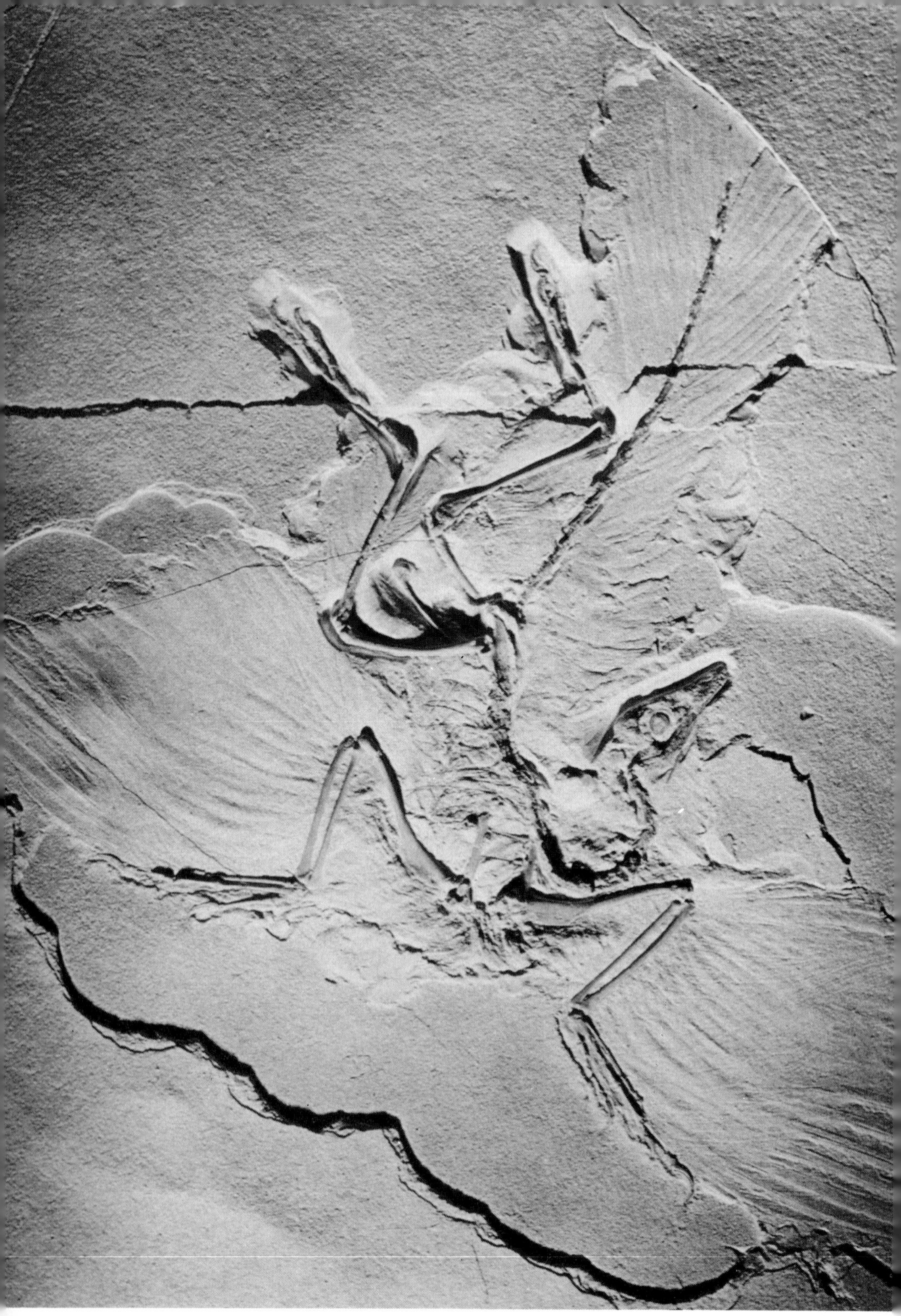

Archaeopteryx lythographica, *the earliest known bird (from Jurassic limestone in Bavaria). Note toothed jaws, grasping claws at base of primaries, and long, jointed tail-bones.*

birds for mastery of the air.

The beginning of the Cretaceous was marked by a surge in the deposition of the calciferous skeletons of microscopic marine organisms, forming the chalk and limestone which gives the period its name. Flowering plants and broad-leaved trees proliferated explosively, giving for the first time a modern air to the ancient landscape. The dinosaurs evolved some of their largest and most impressive forms, including the herbivorous, triple-horned Triceratops, which bore a striking resemblance to an outsize rhinoceros, and the eighteen-foot-tall Tyrannosaurus rex, the largest land carnivore the world has ever seen. The long necked marine reptile Plesiosaurus attained a length of fifty feet, while in the skies a host of flying reptiles belonging to several unrelated groups culminated in the gigantic forty-five foot span of Quetzalcoatlus, a Pteranodon-like creature the size of a small aeroplane.

The end of the Cretaceous was marked by the mass extinction of nearly all the large reptiles and many of the smaller ones. The extinction of the dinosaurs did not happen overnight but probably took place over a period of about a million years, which in geological terms is quite sudden. Mammals had remained small and insignificant creatures throughout the whole of the Jurassic and Cretaceous, but now they swiftly evolved into new and remarkable forms to take the place of the vanishing reptiles.

Whether the rise of mammals had anything to do with the final demise of the giant dinosaurs about sixty-five million years ago is a matter for speculation, but it seems unlikely that such puny shrew-like creatures could have provided sufficiently intense competition, despite their greater activity and slightly higher intelligence. The extinction of the dinosaurs therefore remains one of nature's great unsolved mysteries, despite many ingenious attempts at explanation. The most plausible of these is some sort of climatic change, but the great stumbling block of this and other theories is in explaining why some reptiles were affected and others not. Many marine and fresh water species died out, yet almost equally large crocodiles and turtles have survived to the present day. The sex ratio of many reptiles is critically dependent on the incubation temperature of their eggs and a change of only a few degrees may sometimes bring about a large imbalance in the sex ratio of hatchlings. In the case of land reptiles possibly some interruption of the food chain could have produced a domino effect, involving many species at the same time. The true explanation probably involves elements of several theories, but we may never know for certain.

The small insectivores which had watched the dinosaurs blunder to extinction were mainly if not entirely marsupials or pouched mammals, which today survive only in Australasia and South America. The supercontinent of Pangaea, divided into Laurasia and Gondwanaland by the Tethys Sea, had begun to break up in Jurassic times and the continents drifted apart. Further massive extinctions were to take place as Antarctica drifted into polar latitudes, bearing its entire cargo of trees and animals to their doom.

By the start of the Cenozoic (usually divided into Tertiary and Quaternary), primitive shrew-like placental mammals had evolved, and these underwent an extraordinary adaptation and transformation to produce such diverse forms as bats, hoofed mammals (ungulates), carnivores and whales. Australia had been set adrift carrying only marsupials, and was not invaded by placental mammals until modern times. Outside Australasia competition with placental mammals has rendered marsupials extinct, except for a few which survive in South America.

One of the first of the newly evolved placental mammals was an arborial or tree-living form not unlike the tree-shrews which are found at the present time in South East Asia. The nimble and acrobatic tree-shrews soon developed grasping hands in order to clamber about the branches and leap from tree to tree. Judgement of distance is essential to this way of life, and to this end the muzzle became foreshortened and the eyes migrated to the front of the skull to give excellent stereoscopic vision, resulting in a lemur or tarsier-like animal—the world's first primate. (The need to judge distance precisely by means of stereoscopic vision has produced parallel modifications in the heads of birds of prey.)

Anyone who has ever attempted to live off the jungle without modern aids and equipment will readily appreciate that this demands considerable knowledge, experience and skill. The jungle is bountiful to those who know its ways, but tends to be harsh on those who do not. It is not surprising, therefore, that intelligence was at a premium in these first primitive primates, far more so than in their more specialised relatives which foraged for plants or insects on the ground. The diet of primates was insectivorous at first, like that of the tree-shrews, but became less specialised as time went by, and they increasingly lived by their wits. Those with larger brains were able to exploit their environment more fully and prospered as a result.

Thus it was that monkeys evolved from lemur-like ancestors and became the predominant creatures of the tropical and sub-tropical forest. The less intelligent lemurs could not compete and consequently fell by the wayside except in Madagascar where, isolated from monkeys by continental drift, they continue to flourish up the present day. Tarsiers and similar prosimians have managed to survive in parts of Africa and Asia, perhaps partly because of their nocturnal or crepuscular habits.

Monkeys became increasingly shorter muzzled, more intelligent, and shorter tailed, and even at this early stage were finding that their grasping hands, though developed for quite another purpose, could be useful servants of the brain and able to carry out a variety of tasks which could not be done by the more specialised fore-limbs of other animals. And so it was that in Africa, about 30 million years ago, the first apes came into being—larger bodied and larger brained versions of their monkey ancestors in which the tail, or outward signs of it, had completely disappeared.

These early apes retained the more generalised limb structure of their monkey ancestors, unlike modern apes which have become specialised for brachiation (swinging through the trees). They were therefore slightly shorter in the arm and longer in the leg, having a more useful and dexterous hand structure with an opposable thumb, but in most other respects were similar to chimpanzees. Like modern apes they were almost exclusively arborial, and soon evolved into the group known as dryopithecines (which includes the genus Proconsul, now more properly Dryopithecus).

Their tree-dwelling way of life meant that a great deal of habitat exploited by more dim-witted creatures was denied to them. Probably their high intelligence did not allow this state of affairs to continue for very long, and about 18 million years ago a more enterprising group of dryopithecines began to spend more and more time on the ground, making longer and longer forays outside the forest into the grasslands and savannahs. It was here they parted from the tree-loving ancestors of the anthropoid apes and went their own way, soon becoming ground-dwelling apes of the plains and grasslands, evolving into a new genus—Ramapithecus. This was a significant parting of the ways, for it marks the point where the ground-living hominids ancestral to man separated from the tree-dwelling pongid line of descent which was to lead to chimpanzees, orangs, and gorillas (Pongidae).

One of the reasons these ground-dwelling apes developed the bipedal stance which had only occasionally been used by their tree-living ancestors was to raise themselves above the level of tall grasses and increase their range of vision. Initially this was mainly to avoid predators. It also had a defensive role by increasing apparent body size to deter aggressors (a common ploy in nature) and freed the hands for defence. Later on as these apes became potent predators themselves, it enabled them to look out for potential prey. The first of the hominid line to habitually stand upright was probably a ramapithecine. It is difficult to say exactly when this event took place owing to the fragmentary nature of the fossil evidence, but indications are that it may have been as early as 12 million years ago.

This increasing reliance on bipedal posture also had another beneficial effect in freeing the hands to become full time servants of the brain. The arms, now no longer needed for swinging through the tree tops, became shorter. The huge neck muscles required by the pongid apes to hold their jutting heads were no longer needed, as the skull could now be balanced on a vertical spine articulating from below rather than at the rear. In response to the need for greater speed on the ground, both to escape predators and to overtake prey, the legs became longer and there were pelvic modifications. The foot lost its grasping function producing, by about ten million years ago, an ape of upright stature with remarkably human body, legs, and feet.

Dietary changes and reliance on improvised weapons for defence reduced the size of the canine teeth, producing even at this early stage a distinctly human pattern of dentition. The ramapithecines included a number of different species which, not surprisingly, tended to become more man-like as millions of years went by, but were all slightly built apes between three and four feet high. Like many other groups of mammals they had a tendency to become larger as they evolved, a trend well seen in the evolution of the horse. They were probably opportunist hunter-gatherers and scavengers, with increasing emphasis on hunting as they grew in size and brain power.

Some amazing discoveries have been made by the Leakeys, working mainly at Olduvai Gorge in Tanzania, and other investigators elsewhere in Africa, which support the conclusion that human body and limb structure preceded human mental development by millions of years. It is probable that the upright hunting apes of around 10,000,000 years ago, their hands freed from their former locomotory function, were already able to make primitive tools and weapons of a simple and improvised character. Indeed, such behaviour is not entirely unknown in modern chimpanzees.

By about 5,000,000 years ago the most advanced of these hunting apes of the African plains had become sufficiently differentiated to form a new group, the australopithecines. By virtue of their intelligence and habit of hunting in groups, they became some of the most feared and efficient predators on Earth. Comparatively lightly built (though this varied from species to species), between four and five feet tall, probably covered with coarse hair but with body and limbs almost human in form, these apes had brains about twenty percent larger than modern apes of the same size and body weight. It seems likely that the smallest species, the gracile australopithecines (typified by *Australopithecus africanus*), were the most active and efficient predators despite their slender build. That small size need be no handicap is shown by the modern Cape hunting dog, which though smaller and less intelligent than Australopithecus, is nevertheless able to kill a zebra.

The instinctive dread of man which is even shared by the big cats probably dates from around this time. Those without dread, even the ancestors of present day lions and tigers, often ended up dead. (A tiger confronted in the wild by a man who stands his ground will almost invariably turn tail, though physically an unarmed man is no match for the big cat. The exceptions are when the animal is startled at close range with no time to retreat, when it has cubs to defend, or—in rare cases of confirmed man-eaters—when old or injured and unable to hunt, driven by hunger to attack easy prey of species normally avoided.) Though only slightly larger in brain-size, Australopithecus africanus had substantially more intelligence than a chimpanzee (an animal which in several respects it superficially resembled). It was certainly able to improvise simple tools, and very likely made some of the primitive stone implements found in abundance at various places in Southern Africa, though these are usually assigned to a later genus. It may even have been able to speak, which would not be surprising in view of the accomplishments of substantially less human modern chimpanzees. Unfortunately we shall never know for certain.

About 3,000,000 years ago there were several species of these humanoid hunting apes living side by side in Southern Africa, though perhaps ecologically separated by different food preferences and life styles. As well as the gracile australopithecines already mentioned (which are thought to have been man's direct ancestors) there were at least two robust species, slightly larger and more brutish in appearance but still very man-like. A tooth and jaw fragment of a robust australopithecine has been found as far afield as Java (at Kedung Berubus).

It used to be thought that Asia was the cradle of mankind, but recent evidence strongly suggests that, as predicted by Darwin, all significant hominid advances originated in Africa and from there spread out to other warm parts of the world. As each successful new design was evolved in the African hominid 'factory' it expanded like a wave through favourable habitat, but probably migration as such was not involved. An average advance of only a few hundred yards a year would reach the furthest corners of South East Asia in about 25,000 years, the bat of an eye-lid by geological standards. Ramapithecus in particular seems to have been abundant in India, where it was first discovered, and several later hominids are known to have reached South East Asia. Australia and the Americas were isolated by continental drift in comparatively early times and were not reached by hominids until invaded by modern man a few tens of thousands of years ago.

Though denied human status by anthropologists and taxonomists because of its small brain (500–550 c.c. as compared to about 1300 in a modern man of the same size and weight), Australopithecus would certainly have caused a furore if it had survived to modern times. Would it be moral for an animal which could speak in human fashion and make tools to be kept in a zoo? Would its employment in doing the many tasks for which its intelligence would have suited it have been slavery and exploitation, or domestication? Fortunately these are questions which we will never have to answer, as the animal appears to have become extinct about a million years ago.

If Australopithecus africanus was not man's immediate ancestor then he was certainly not far removed. About two and a quarter million years ago, perhaps even earlier, a larger brained species developed from an australopithecine ancestor, and this was the first recognised species of man, with a cranial capacity of around 750 c.c.—about half that of modern man.

This early species of man was discovered by Mary Leakey in 1960 at an excavation in Olduvai Gorge, Tanzania, and has been provisionally named *Homo habilis.* A much better preserved specimen was found in 1972 by her son, Dr. Richard Leakey, at Lake Turkana, Kenya. Almost certainly this was a direct ancestor of modern man, and it is to this species that the earliest known stone tools (dating between two and three million years old) are usually attributed. It is slightly misleading to talk of *the* missing link, as several such links are required to bridge the gap between apes and modern man, but of all the candidates for the title the nicest balance of human and ape characteristics is probably Homo habilis. Mid way in stature between pygmy-sized Australopithecus and modern man, this early ape-man combined a hairy but human body with an ape-like physiognomy and a brain size about fifty percent larger than in modern apes of similar body weight.

The next important advance came almost two million years ago with the appearance of Homo erectus, also known as Java man (formerly Pithecanthropus erectus, but now Homo) and Pekin man (*Homo erectus pekinensis*). Though originally discovered in the Far East, much earlier examples of Homo erectus dating back more than one and a half million years have since been found in Africa. This is in agreement with the idea of successive waves of increasingly advanced hominids radiating outwards from their African birthplace to the rest of the Old World tropics. A later example can be seen to this very day in the Negrito aborigines of obviously African origins which are found in the jungles of West Malaysia and the Philippines.

With his sloping forehead, prominent brows and receding chin, the brain capacity of Homo erectus varied from one individual to another over a period of about one and a half million years, but averaged around 950 c.c. As might be expected, the earliest examples tend to have slightly smaller craniums while the largest are found in more recent fossils, especially those from China (Pekin man, c. 500,000 years B.P.), some of which have cranial capacities of around 1 200 c.c.

Then about a quarter of a million years ago came Homo sapiens neanderthalensis, formerly regarded as a separate species (*Homo neanderthalensis*) but now considered by most authorities as belonging to the same species as ourselves and given sub-specific status. A similar argument rages over the validity of the Homo status of Homo habilis, regarded by some authorities as Australopithecus habilis. Problems like this occur quite often in taxonomy, and not surprisingly. There must obviously come a time when a species or sub-species is about mid way in the transition from one clearly defined evolutionary stage to another when both points of view are equally correct, or as near equally as makes no odds.

This brings us almost up to modern times, as the cranial capacities of the earliest examples of Neanderthal man are around 1100 c.c., ranging to about 1350 c.c. in some of the most recent individuals. These had brains comparable in size with that of modern man, but robust physique and coarse features still slightly reminiscent of an ape. Neanderthal man can be divided into two groups according to the age of the fossils: primitive Neanderthal which seems to have originated in Africa, and advanced Neanderthal which occupied Eurasia until the middle of the last ice age. Primitive Neanderthal, as might be expected, looks very similar to advanced Homo erectus from which it can barely be distinguished by the inexpert eye, and is probably the form from which our own subspecies (*Homo sapiens sapiens*) is descended. So far as can be ascertained at present advanced Neanderthals appear to have died out about thirty to forty thousand years ago, perhaps exterminated by the modern men who took their place.

Wherever modern man came from one thing is certain: he did not evolve suddenly 40,000 years ago in a genetic quantum leap, but moved into Europe from elsewhere, probably from Asia. The fact that Neanderthal man disappeared at about the same time can be explained in mundane terms which are entirely consistent with what we know about human nature. Divergence from a common stock took place very much earlier, and in view of the previous history of hominid innovation an African origin for modern man is very much on the cards.

It is thought by some that Neanderthal man never died out completely and that some of his genes live on in modern man. It is supposed that when the inferior Neanderthals were attacked and decimated in wars with modern man, the victors interbred and intermarried with some of their female captives. If prehistoric man was anything like as lecherous as ourselves this seems very probable. It may be that the more robust build and coarser features of certain races (including ourselves) as compared with, say, orientals, is because of the number of Neanderthal genes in their genetic make-up. This theory is highly speculative and there may be nothing in it, but among other things it would nicely account for the slightly Neanderthal features of Australian aborigines. Some fossil skulls from the Eastern Mediterranean which seem to be transitional between Neanderthals and modern man can easily be explained in terms of interbreeding.

Perhaps one day a perfectly preserved specimen of Neanderthal man may be found frozen in some glacial crevasse or entombed in the tundra permafrost, like the carcases of mammoths which have already been found. That such specimens exist is almost certain, as Neanderthals were fairly numerous forty thousand years ago, contemporary with mammoths which they hunted, and man is naturally predisposed towards exploring hostile terrain and as a consequence getting himself into difficulties. The chances of such a specimen being found, let alone recognised for what it was, are, however, remote. For every mammoth carcase found, a hundred others rot in the Arctic summer sunshine or are carried to the sea in some Siberian river. The alleged Neanderthal man entombed in a block of ice and exhibited at fair-grounds of recent years was almost certainly the corpse of a modern orang utan.

If you had thought the similarity between ourselves and chimpanzees was merely a coincidence, you will now see how wrong you were. The geneological records at Somerset House go back only a few hundred years, but if we could research such records back for eighteen million years or so we would then find documentary evidence that we shared an ancestor with that very same chimpanzee you saw on your last visit to the zoo. The chimp in turn would find that a few million years ago he (and therefore, more remotely, we) shared a distant grandfather with the gorilla or orang in the neighbouring cage.

If we could compress the entire four billion year history of life on Earth into a mere twenty-four hour day, then on this scale our parting from the ancestors of modern anthropoid apes would have taken place

†The 4–5 million year protein electrophoresis date for hominid separation is in conflict with the fossil evidence and will be discredited.

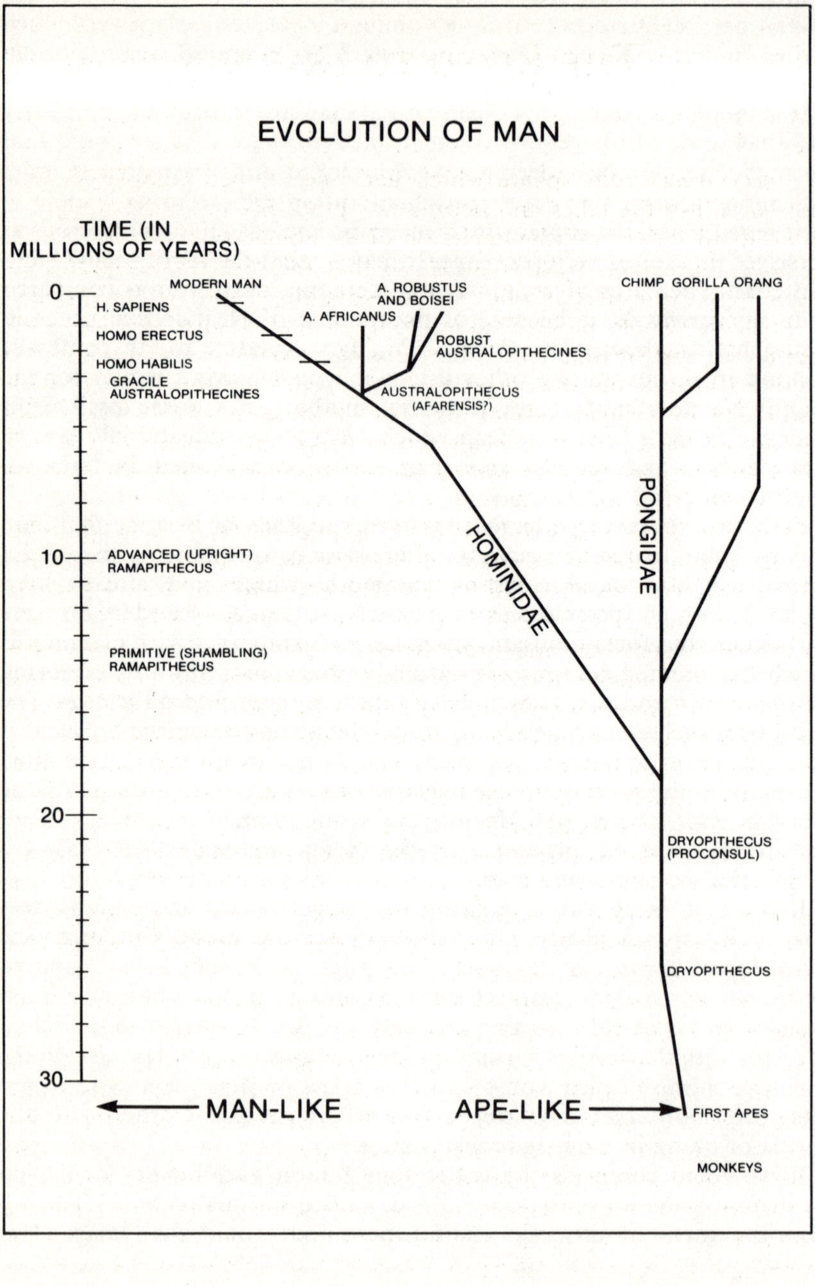
EVOLUTION OF MAN
TIME (IN MILLIONS OF YEARS)
0
10
20
30
H. SAPIENS
HOMO ERECTUS
HOMO HABILIS
GRACILE AUSTRALOPITHECINES
ADVANCED (UPRIGHT) RAMAPITHECUS
PRIMITIVE (SHAMBLING) RAMAPITHECUS
MODERN MAN
A. AFRICANUS
A. ROBUSTUS AND BOISEI
ROBUST AUSTRALOPITHECINES
AUSTRALOPITHECUS (AFARENSIS?)
HOMINIDAE
CHIMP
GORILLA
ORANG
PONGIDAE
DRYOPITHECUS (PROCONSUL)
DRYOPITHECUS
FIRST APES
MONKEYS
MAN-LIKE
APE-LIKE

only about five minutes ago. So 18 million years,† though a long time in terms of human lifetimes, is really quite recent. You shared an ancestor with your Christmas turkey back in the Permian nearly 280 million years ago, and your ancestor in the Devonian was closely related to a distant cousin which you ate with your last fish and chips. More amazing still, if detailed records went back three billion years to the early Precambrian you would find you shared a very distant grandfather with the apple trees or rose bushes in the back garden!

Curiously enough, this evolutionary sequence is mimicked by the development of the human embryo, which begins with a fertilised egg-cell strongly resembling a unicellular amoeba-like ancestor from which we parted company somewhere in the middle Precambrian. This tiny cell then rapidly multiplies by binary fission into a colony of identical cells. These then differentiate to form organs, one of the first of which is the notochord, a primitive backbone found in our sea squirt and lancelet ancestors. The developing embryo closely parallels that of a fish for a few days, having apertures bearing a remarkable resemblance to the gillslits of sharks and rays.

Beyond this stage comparisons with the embryos of both reptiles and birds (which were not only our cousins but our brothers about 280 million years ago) show a striking similarity, even in the development of tails. Again we part company, and for a while the human embryo closely parallels the development of other mammals, firstly shrews then monkeys and apes. At about 18 weeks there is an amazing similarity between the human foetus and those of apes, including a covering of fine fur. We then part company with our animal forebears and the foetus begins to take on an increasingly human appearance.

Some will say that these resemblances are merely superficial and without real significance. I would agree to some extent with the first part of this statement, for however similar the embryo may be at a certain stage to that of a reptile, it is not a reptile and will go on to become a human being, which an embryo crocodile, for example, will never do. Nevertheless, to write the whole thing off as a mere coincidence seems to be bending the facts to breaking point, a deliberate evasion of the truth.

There is little doubt that in the development of the human foetus and its resemblance, however superficial, to evolutionary stages of our development, we are watching a re-enactment of the evolutionary drama which has its roots in biological history and owes no more to coincidence than does our resemblance to the anthropoid apes. Our observations could perhaps have been started a stage earlier, for a segment of the pre-embryonic DNA resembles a primeval macromolecular ancestor which drifted in the chemical 'soup' of the early Precambrian where our story began, and this too is not a coincidence.

But do not get too worried about eating or otherwise exploiting your distant cousins—they have no hesitation in doing the same to us when they think they can get away with it. Indeed, some of our most deadly enemies are bacteria and viruses, distant cousins from four billion years ago. We should be careful, however, not to drive any of our relatives into extinction (with the possible exception of few deadly microscopic enemies), as they represent a priceless heritage which we are in danger of frittering away. Apart from anything else, the story of life is written in almost full detail by an amazing sequence of living fossils which have managed to survive into our own time. Once lost they are gone forever.

These chapters have of necessity been slightly technical, but were necessary in order to prepare you for what comes next. I have tried to avoid going into unnecessary detail, but if the exciting story of reality seemed dull or dry at times the fault was mine. Just to make sure that you have sufficiently understood it, let us recapitulate the salient facts before we move on:

There is no such thing as absolute motion or absolute rest and velocity can only exist relative to some object or objects forming a frame of reference. All forms of energy are inter-convertible—any one may be changed into another (heat into electrical or mechanical energy, for example, or chemical energy into heat and light). Mass and energy are but two different forms of the same thing (if this was not true atomic bombs

VENUS

95 % CO_2 (carbon dioxide)
3·5 % N_2 (nitrogen)
1 % CO (carbon monoxide)
0·01 % Ar (argon)
0·1 % O_2 (oxygen)
0·4 % others

JUPITER

85 % H_2 (hydrogen)
11 % He (helium)
1·5 % CH_4 (methane)
1·5 % NH_3 (ammonia)
0·5 % N_2 (nitrogen)
0·5 % others

EARTH

78 % N_2
21 % O_2
1 % Ar
0·0325 % CO_2
0·00182 % Neon
0·000524 % helium
0·00015 % methane
0·000114 % krypton
0·00005 % hydrogen

SATURN

Similar to Jupiter

TITAN (Moon of Saturn)

nitrogen 85 %?
methane 10 %?
hydrogen 2 %?
argon 1 %?
others 2 %?
(ammonia and CO_2 probably frozen out as ices)

MARS

95·4 % CO_2 (carbon dioxide)
3 % N_2 (nitrogen)
0·3 % O_2 (oxygen)
0·16 % CO (carbon monoxide)
1 % Ar (argon)
0·1 % others

URANUS

hydrogen 79 %?
methane 10 %?
helium 10 %?
others 1 %?

NEPTUNE

Similar to Uranus

ATMOSPHERES OF THE PLANETS

The variations are due mainly to mass (gravitation) and heat (proximity to the sun). Also, in the case of Earth, to modification by living organisms. Figures for water vapour, sometimes an important constituent, have been omitted. Amounts vary not only from place to place but in the same place at different times.

The nearest in composition to the primordial nebula is the atmosphere of Jupiter, which has retained light gases because of its large mass and low temperature. Figures for Earth and Mars are exact, Jupiter, Saturn and Venus approximate, and other planets only a rough approximation. Water vapour is in short supply on Venus and Mars, but there is evidence that Mars once had a much denser and moister atmosphere than it does at present.

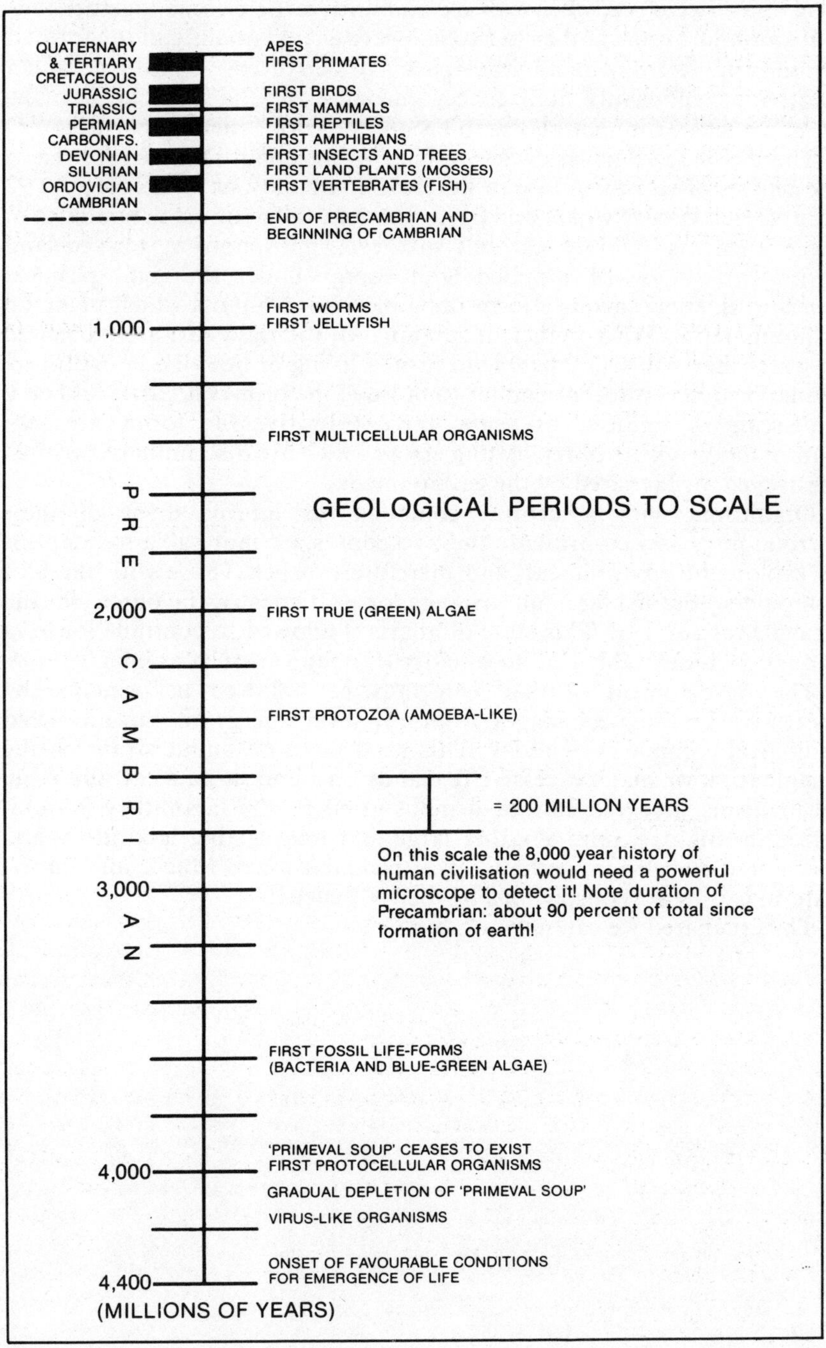
QUATERNARY & TERTIARY
CRETACEOUS
JURASSIC
TRIASSIC
PERMIAN
CARBONIFS.
DEVONIAN
SILURIAN
ORDOVICIAN
CAMBRIAN
APES
FIRST PRIMATES
FIRST BIRDS
FIRST MAMMALS
FIRST REPTILES
FIRST AMPHIBIANS
FIRST INSECTS AND TREES
FIRST LAND PLANTS (MOSSES)
FIRST VERTEBRATES (FISH)
END OF PRECAMBRIAN AND BEGINNING OF CAMBRIAN
FIRST WORMS
FIRST JELLYFISH
1,000
FIRST MULTICELLULAR ORGANISMS
GEOLOGICAL PERIODS TO SCALE
PRECAMBRIAN
2,000
FIRST TRUE (GREEN) ALGAE
FIRST PROTOZOA (AMOEBA-LIKE)
= 200 MILLION YEARS
On this scale the 8,000 year history of human civilisation would need a powerful microscope to detect it! Note duration of Precambrian: about 90 percent of total since formation of earth!
3,000
FIRST FOSSIL LIFE-FORMS (BACTERIA AND BLUE-GREEN ALGAE)
'PRIMEVAL SOUP' CEASES TO EXIST
FIRST PROTOCELLULAR ORGANISMS
4,000
GRADUAL DEPLETION OF 'PRIMEVAL SOUP'
VIRUS-LIKE ORGANISMS
ONSET OF FAVOURABLE CONDITIONS FOR EMERGENCE OF LIFE
4,400
(MILLIONS OF YEARS)

would not work and the sun and most other stars would have gone out long ago). Therefore, when particle and anti-particle come together they will annihilate one another to become pure energy (mainly in the form of photons or electromagnetic wave packets). The reverse is also true, and a photon of sufficiently high energy encountering the force fields of an atomic nucleus can materialise to become a pair of charged particles.

These charged particles, when subject to the conditions which exist at the centres of stars, can fuse in different combinations to become the 92 natural elements of which our Earth is made. Some of the commonest of these elements (mainly carbon, nitrogen, phosphorous, oxygen and hydrogen) can, when supplied with energy under the right physical conditions, spontaneously form the molecular sub-units which make up living material. When sufficient sub-units of the right sort are incubated in water they will self-assemble to form a living or near-living organism. Under benign conditions similar to those of the primeval Earth, and on a time-scale of millions of years, these primitive life forms are self-improving by virtue of replicating errors which are discriminated against (destroyed) or favoured by the environment.

Organisms with a central computer to inform them of their surroundings and co-ordinate their responses are more able to adapt to or exploit the environment, and therefore prosper. Those with the best computers (brains) are, on average or statistically, favoured in the struggle for survival. This state of affairs if allowed to continue for long enough will inevitably lead to intelligent organisms such as man.

The development of man can thus be followed in principle by successive and logical stages from primeval energy to the advanced biological computers of today. This need place no undue strain on the imagination or make excessive demands on blind faith, as all the basic mechanisms involved can be demonstrated in the laboratory (natural selection might require a rather large and long lasting laboratory and thermonuclear fusion perhaps an expendable one). Magic and supernatural causes are conspicuous by their absence.

Thus prepared we can now move on.

IV

Savage Shores

"Primitive people, to whom many aspects of the physical world are strange and terrifying (because not understood), fill their environment with imaginary gods, demons, ghosts and monsters."—Isaac Asimov

Owing to the power of suggestion, the herd mentality, and the familiarity which breeds acceptance, some most extraordinary and irrational ideas will seem quite reasonable when enough of the herd believe in them. The grotesque extremes of fashion, including the self-mutilation practised in primitive societies, are a mild example of this. Conversely, some quite reasonable ideas such as cremation of the dead, the abolition of slavery, religious freedom, freedom of speech and the press, and universal suffrage were considered unreasonable or even wicked until adopted by the herd.

This herd mentality is the age old way of the wild which it is now time to cast aside, and whilst we must acknowledge the reality of our inheritance from Australopithecus, we should try to imitate him as little as possible and encourage traits which will emphasise our separation rather than our similarities. Our human-ness increases as we diverge from our primitive ancestors, rather than being embodied in the animal traits so esteemed by popular belief. It is for this reason that a prominent palaeo-anthropologist has suggested, only half jokingly, that *we* are the missing link between apes and man!

There is a small island in the Atlantic where, despite centuries of British rule and in defiance of all efforts by the authorities to educate them, the natives still live in a primitive world of ignorance and superstition. Even today, years after the introduction of compulsory education, many fail to grasp simple facts which in our country it is popularly supposed that every schoolboy knows (what coal is or what rainbows are, for instance, or why the sun is hot), let alone such comparatively weighty matters as what stars are, or atoms, and the workings of evolution and DNA. The majority have little knowledge of events in the outside world, which might as well be on another planet as far as most of them are concerned. But what does it matter if a few natives are superstitious and educationally backward, you may be wondering, even if they are a British responsibility? It matters a lot, as will soon become clear.

Discounting a purely racist explanation of this unhappy state of affairs, some observers blame the British-run education system. Others attribute blame to the primitive beliefs and superstitions which flourish in abundance, creating an atmosphere hostile to new ideas. Most of the natives believe in magic in one form or another—anything from astrology to animism and witchcraft. One was recently beaten to death by his companions to drive out evil spirits. The use of drugs to produce a tranquil or euphoric state of mind untroubled by reality is widespread.

The adoption of Western dress by the majority of the population is deceptive, as it disguises attitudes more readily associated with grass skirts and bones stuck through the nostrils. Despite a thriving tourist trade there are still many parts of the island where it is dangerous for travellers to venture, especially alone or at night, for fear of attack by savages. Much time is taken up by the regular fertility dances which are a feature of community life and seem to be based on the courtship displays of some of the island's birds, possibly grebes or cranes. They are particularly reminiscent of the antics of certain birds-of-paradise which are, however, not native to the island. On the northern coast one of the last great auks, a now extinct diving bird, was unfortunately clubbed to death as a witch by the local inhabitants.

The name of this strange island, if it had not already dawned on you, is Great Britain. And while I have obviously written these paragraphs tongue-in-cheek they are neither fiction nor gross exaggeration, as you can easily check for yourself. Admittedly, it was back in the 1840s that one of the last great auks was mistaken for a witch, but someone was beaten to death at an exorcism as recently as 1980, so times have not changed very much. The natives of 'perfidious Albion' (as the island is sometimes called) are internationally renowned for their anachronistic dedication to the idea of class, the self-inflicted injuries of their ailing industries and the hysterical excitability of their football hooligans.

Researches have shown that the average person's grasp of reality is little better than that of a cow, a rabbit, or a dog, or at best a New Guinea tribesman. Like Australopithecus, they are only in touch with the real world in so far as is necessary for them to go about their business in a reasonably workmanlike fashion. "Nonsense!" I can almost hear you say. Is it? Consider the beasts of the field, which have watched the stars come out on every clear night throughout their lives without ever once pausing to wonder what they were looking at. The only reality they know is the superficial reality of their own immediate surroundings and their only interests are their material needs and their interactions and relationships and others of their species. Most human beings are much the same.

If you think the comparison far-fetched, ask a few of your friends what those twinkling lights might be that they have seen so many times in the night sky. Nearly forty years into the space age the sort of answers you will get are "I don't know" or "Aren't they sort of—er—planets?" On an average night only two or three will be planets. Some will say "Stars, of course!", but before getting too excited at the prospect of someone who is aware of the world about him, ask them what a star is. Astonishingly, most people do not know! On a recent television current events programme, about half-a-dozen people, ostensibly chosen at random, were asked what uranium was. The only one who gave a sensible answer turned out to have been briefed before-hand by the interviewer. And this in an area where it was proposed to mine uranium, with all the attendant implications of a health hazard to local people. The local cows were similarly ignorant of the problem, but they were not interviewed.

Further enquiries will reveal that this is just the tip of the iceberg and that forty years into the Atomic Age the majority do not know what an atom is or have such important and basic information as a knowledge of evolution, and like the dog or the cow have never even wondered about it! It is not a question of intellectual snobbery or one-upmanship, as no feats of intellect need necessarily be involved, and there may often be quite practical applications of a sound knowledge of the world we live in. A Ph.D. in physics or biology is not required in order to understand the basic details of such things—on the contrary, the average urinal attendant or brickies' labourer (perhaps even a bright chimpanzee) can readily understand the basic concepts once they have been properly explained. It would scarcely be too fanciful to say that the fateful day in the remote past when our ape-like ancestor first wondered what the stars were marked our transition from animal to human.

At least animals have an excuse—they are not provided with sufficient grey matter. Even the New Guinea tribesman can justifiably claim that he is not provided with sufficient education. What reasonable excuse there might be for civilised men is a problem demanding the sort of unseemly mental gymnastics and intellectual contortions sometimes known as 'doublethink'—the same doublethink that regards honesty and truthfulness as unwholesome when the real unwholesomeness lies rather nearer home. It is not enough for civilised man to limit his perception of reality to the superficial aspects of his own immediate surroundings. That is an attribute of animals. All human progress has been a search for truth—it is man's destiny to wonder and enquire. For better or for worse we began our journey along the road that first fateful day when our ancestor wondered about the stars. There can be no turning back.

Let us examine the average man's perception of reality. The astonishing truth in the final quarter of the twentieth century is that the average Briton still believes in some form of magic! They may not actually call it magic—there are a variety of popular euphemisms. More likely they will speak of witchcraft, the paranormal, astrology, or miracles, but in essence what they are referring to is magic, whatever they might choose to call it.

The various irrational beliefs and primitive superstitions widely held by the 'natives' may be divided into two main categories: those which appertain to magic and the supernatural, and those which do not. We will first deal with irrationalities of a magical nature, which may be further subdivided into religious and non-religious.

It was very tempting to be like most other writers and dodge the issue on religion, as this prickly subject is the type species of the irrational ideas programmed into the sub-conscious of almost every individual, and nothing is more likely to cause resentment when roughly handled. But honesty compels me to grasp the nettle. Besides, any book on reality (let alone one on unmentionable subjects) which failed to do so might fairly be prosecuted under the Trades Descriptions Act. (Dr. Isaac Asimov, the well-known scientist and author, has an amusing way of dealing with the situation when asked whether he believes in God or religion. In order not to risk giving offence, yet at the same time to avoid telling a lie, his stock answer is "Whose?"!)

To make wickedness out of virtue and virtue out of wickedness, to suppress truth and reward the guilty, has so often been the effect if not the intention of religion through the ages that examples are not hard to find and you will not need me to spell them out for you. The ideas I criticise are not harmless, as will become all too chillingly apparent later in the book. Religion's new found enthusiasm for truth and freedom is in stark contrast to the historical record. The leopard has changed its spots, they would have us believe. But are the new markings only camouflage, and could it be that, underneath, the leopard is just the same as it always was? Does the new enthusiasm for truth amount to no more than lip service? Time will tell. To see religion in true perspective it is probably best to look at those not widely followed by the natives of our own particular island. It is always easier and less uncomfortable to see fault in others rather than in ourselves.

The divinity of kings was widely believed in Japan and Tibet until quite recently. Under the Shinto religion of Japan, the Emperor Hirohito was regarded as a god until 1946, and to die for him, especially as a 'kamikaze' (divine wind) suicide pilot, was believed to ensure a place in paradise. Can 70 million Japanese be wrong? The short but perhaps unpalatable answer is—yes, they can.

Buddhism has some good ideas such as the way of gentleness and acceptance of the impermanence of material things, but it also has the Dalai Lama, the so-called god-king of Tibet, and preaches belief in reincarnation. Harmless enough, you might think. A Minister of Education in an Eastern country is said to have replied complacently to complaints about the death of children from disease: "After all, they will be reincarnated anyway!" To the best of my knowledge the story is true and not allegorical. Incidentally, both Emperor Hirohito and the Dalai Lama are still alive at time of writing.

The Rastafarians of Britain and the West Indies believe the late Emperor Haile Selassie of Ethiopa to be a god. Tyrant and despot he may have been, but god . . .? A number of other people, not all of them royalty, have proclaimed themselves or been proclaimed by their followers as gods in the present century. In the Solomons an American named John Frumm has been deified by the natives. Statues are made to him, with models of the aeroplanes, radio masts and airfields which will supposedly bring him back to the islands bearing limitless 'cargo' for free distribution to the inhabitants. Ceremonies are performed by priests dressed in imitation American uniforms and some adherants of this 'Cargo Cult' maintain that John Frumm lives in a volcano! The convention that any nonsense, however ridiculous, becomes respectable when you call it religion is plainly absurd and must be abandoned.

The Hindu faith has almost innumerable gods, among the main ones being Vishnu, Shakti, and Shiva. No wonder Dr. Asimov has to ask "Whose?"! According to the scriptures these gods get up to all manner of unlikely adventures, but some interpretations hold that these stories are allegorical and that all the hundreds of gods and their wives are but different manifestations of one god, Brahma. Hinduism is therefore very elastic and can be all things to all men. This is the religion of the notorious sacred cows, which have been the cause of much bloodshed through religious disputes and live in pampered luxury while their worshippers often starve. The insane idea of 'untouchables' originates in the Hindu caste system, and if it strikes you as perfectly reasonable it is a pretty safe bet that you are not an untouchable, hundreds of whom have recently been murdered in religious riots. Both Buddhists and Hindus believe in reincarnation, but not necessarily reincarnation as humans. It seems we may be born again as snails, insects or similar lowly creatures!

It is easy to see the absurdity of these foreign religions as we are not closely involved, but nearer to home it is another story, and we often cannot see the wood for the trees. A surprising number of people still believe in the Biblical Flood in spite of all geological evidence to the contrary. A tsunami or tidal wave from the explosion of Thera, a volcanic island off Crete, undoubtedly inundated coastal areas of the eastern Mediterranean at about the relevant time (1500 B.C.) in a stupendous Krakatoa-style eruption which was probably the source of the legend, but a universal flood is quite another matter.

These people fail to realise that water can only be added to one place by taking it away from another, and there is not enough water locked up in the polar icecaps to flood the entire Earth. If the water of one hemisphere was used to flood the other, there could by definition have been no universal flood. Moreover, in a universally flooded world there would have been nowhere for the waters to drain away, and the very fact that by their own admission the flood receded is proof that it was able to drain away into areas that were not flooded. Such basic logic makes even the geological record seem superfluous.

Another point which seems somehow to have escaped attention is that all the navies ever built would scarcely have sufficed to carry a viable breeding nucleus of all the world's animals and their provisions. Certainly no ark could have taken them even if it had been the Ark Royal. Leaving aside the fact that Australia and the Americas had not even been discovered at the time, the task of getting the animals together would have occupied an army of wildlife experts for their entire lives whilst another army of woodcutters and carpenters built a fleet which would have been the eighth wonder of the world. But surely, you may say, nobody believes in Noah's Ark nowadays? An Apollo astronaut has recently organised his second expedition to Mount Ararat to look for it.

The creation of Adam from a model of clay is such an absurd proposition that it is astonishing that anyone could take it seriously in the final quarter of the twentieth century. The fact remains that many people do, ignoring an avalanche of fossil and biological evidence to the contrary. This represents an intellectual feat requiring truly staggering powers of self deception.

The magical transformation of one object into another in defiance of the laws of physics and common-sense is a popular source of irrational superstition. The staff of Moses which supposedly became a snake, water into wine, the bread and wine which are said to become the body and blood of Christ in the Eucharist are well known examples. While the transmutation of elements—hydrogen into helium, uranium into lead, nitrogen into carbon—certainly is possible under certain extreme physical conditions or inordinately long time-scales, the complex molecular transformations listed above are not, except by the sort of unmiraculous biological processes you can witness every day in your own garden.

Experiments have shown that walking on the water without artificial aids is only possible where the water has previously been frozen, unless you happen to be a stormy petrel or one of several other tiny wild creatures small enough to accomplish the feat by making use of surface tension. One of these, a small tropical lizard known appropriately enough as the Jesus lizard, is assisted by large feet and ultra-rapid leg movements.

Another primitive belief often encountered is the belief that the dead can be raised. It should be understood that the body does not die immediately when the heart stops beating. The cells of the various organs shut down in groups at progressive intervals over a period of hours according to their ability to withstand the loss of oxygen and nutrients normally supplied by the blood. The first cells to go are the brain cells, which after about four minutes without oxygen are irreparably damaged. If the blood supply is not quickly restored, brain death ensues a few minutes later. On the other hand the cells of kidneys and certain other tissues are able to remain alive for several hours after the heart has stopped. The heart itself has no special significance—it is only a pump for the circulation of blood.

It follows that people can be brought back from the dead only because they have never really died. Once brain death has occurred resuscitation is no longer possible, despite the fact that certain other tissues may still be living. Irreversible damage occurs in a very short time after the heart stops beating, but provided resuscitation procedures are initiated within this critical period it is possible to produce the illusion of raising someone from the dead.

Belief in life after death is another primitive superstition often held by the 'natives' despite the fact that there is no provision for anything but the physical body incorporated in the DNA blue-print for a new individual. Just whereabouts along the road of evolution this 'soul' or spiritual body appeared is something which believers are at a loss to explain. It is generally held that souls are an attribute exclusive to human beings, which forces us to the strange conclusion that one of our ape-like ancestors with no soul of its own must nevertheless have given birth to offspring with souls. This is entirely inexplicable in terms of evolutionary theory and natural selection, especially as possession of souls would confer no obvious advantage by way of increased reproductive fitness.

The timing, however, is significant. How very strange that souls should suddenly appear just as our ancestors' brains were becoming sufficiently complex to produce that product of intellect and self-awareness, the human mind. Could it be that they are one and the same? Could it be that intellect is not the product of the soul, as some would have it, but the other way around? It could. The mysterious 'soul' is no more than an illusion produced as the by-product of a computer, and its elusive, will-o-the-wisp, mirage-like nature, the ability to keep always tantalisingly just beyond our grasp, is thus explained. The coincidence turns out to be no coincidence at all.

If, on the other hand, you support the idea that souls evolved at an earlier stage or came into being as an integral part of the very first life forms, even more preposterous situations present themselves. Bear in mind that contrary to popular belief the dividing line between living and non-living material is not always a sharp one (in contrast to the boundary between reality and make-believe which should be sharp but in most people is somewhat blurred). The crystalline form of the tobacco mosaic virus, for example, is not alive by most of the standard criteria which define living organisms, yet is capable of infecting plants and initiating the production of succeeding generations of virus.

The dividing line would have been even less well defined in the case of the very first living organisms, our direct ancestors, which were nothing more than self-replicating molecules floating in pools of chemical soup. Their entire physical, spiritual, and evolutionary potential was no more than a consequence of the arrangement of individual atoms. There is no fundamental reason why similar primitive life forms should not be produced in the laboratory, and this will probably be achieved within the lifetimes of people now living. It strains our credulity to the utmost to believe that souls could exist independently of the nucleic acid mechanism which determines the properties of all living things, defying all scientific techniques of analysis and detection, but to believe that such weird entities could be brought into being by means of a chemical reaction must be the ultimate perversion of logic and reason.

A number of other problems cry out for answers if the idea of souls is to be at all credible. Do viruses, fish, or reptiles (modern representatives of primitive ancestors in our evolutionary tree) have souls? If so, how and where did they acquire them and (unless we are to presume that all souls are the same whether virus or ape) by what mechanism did they evolve into the soul of modern man? Belief in life after death would appear to be more a product of wishful thinking than of sound reasoning or logical deduction.

Many people would have us believe that there is a whole category of living things which biologists have failed to include in their classifications and check-lists. I refer to the multitude of angels, seraphs, cherubs, devils and demons in their imaginary kingdoms of heaven and hell from whence they supposedly make occasional sorties to the surface of the Earth.

Small wonder that no self-respecting biologist will have any truck with such nonsense, despite the fact that many people believe in it and the beings concerned can be described in considerable detail, sometimes from 'eye witness' accounts. Certain lunatic fringe religions which shall remain nameless are particularly notable in this respect.

To be fair, there are few religions entirely without any good points or redeeming features at all, but as descriptions of reality most of them are laughable. It is hard to see why such truth as exists cannot be accepted without the claptrap and mumbo-jumbo which usually goes with it. Having grasped the nettle we can now move on to examine irrational beliefs of a magical but non-religious nature. Like mosquitoes in a nudist colony, we are confronted by such a choice of tempting targets that it is a problem to know where to begin.

In Great Britain alone there are many people who believe in so-called 'psychic surgery' which purports to reproduce by magical means the achievements of modern medical science, using nothing but the operator's bare hands. Cancers and tumours are allegedly removed and various other kinds of remedial surgery 'performed'—I use the word advisedly. This service does not come free, though the victims are sometimes asked for donations rather than charged a fee, and hundreds if not thousands of people have fraudulently been relieved of considerable sums of money. But this is not the main objection. The real danger is that people in need of orthodox medical treatment will be tempted to avoid it, sometimes at the peril of their lives, to seek remedies in quack medicine.

The Amazing Randi, world famous magician and illusionist, has devoted a substantial amount of time and effort into exposing the practitioners of this and other forms of the so-called para-normal. Randi, an American who has made several appearances on British television, has reproduced most if not all of these phenomena, including psychic surgery, but lays no claim to psychic or para-normal powers. His message is that make-believe and illusion are for entertainment only and not to be used as substitutes for reality—a welcome voice of sanity. The British magician Paul Daniels, who is also able to reproduce para-normal phenomena ". . . under laboratory conditions", is known to hold similar views. Because these things are due to malfunctions of the electrical circuitary of the brain or to faulty programming rather than any external phenomena, investigation of E.S.P., the supernatural, and much of the so-called para-normal lies in the realm of psychology rather than in more exact scientific disciplines such as physics.

Astrology, the belief that events on Earth are profoundly influenced by the positions of distant stars and planets, is widely practised by people who are apparently unable to grasp the principle of cause and effect. A few celestial bodies undoubtedly can influence events on Earth. The sun, for example, not only makes life on Earth possible but is responsible for the weather and to a lesser extent for the tides. On the one hand we have the sun—the cause. On the other we have our weather—the effect. And in between we have 93 million miles of space. Clearly we need a mechanism which will bridge the void, linking cause with effect. Surprise, surprise, we find several such mechanisms, thus neatly avoiding the need to invoke magic. Streams of electromagnetic waves, sub-atomic particles and the gravitational force link cause with effect just as logic and reason require they should do.

Consider Sir Francis Drake on Plymouth Hoe, playing his historic game of bowls. On being told that the Armada had been sighted in the Channel he nonchalantly decided to continue his game and leave the Spaniards to be dealt with later. Strangely enough this was an astrological effect, the cause of which lay 240,000 miles away in space—the moon as it orbited the Earth. You may be wondering how the moon affected Sir Francis' decision, and may be disappointed to learn that magic was not involved.

Just as before, logic requires us to find a mechanism linking cause with effect. We have it in the gravitational field which reached from the moon to the Earth, causing the tide which would permit the English fleet to sail and which Sir Francis, seaman that he was, knew was not due for a couple more hours, so permitting him to finish his game.

When we examine other celestial bodies, however, astrology ceases to make sense. "When beggars die there are no comets seen—the heavens themselves blaze forth the death of princes" wrote Shakespeare. Our problems arise when we try to link cause—some distant star, comet or planet—with effect: some event or cataclysm upon Earth. The planets have no light of their own and their feeble gravitational effects on us are almost unmeasurable, far less than those of a nearby hill, tower block, or mountain, or deposits of heavy metal ores. Although they produce their own radiations the light and gravitation of stars is similarly weak because of their immense distances, and hopelessly inadequate to explain alleged astrological effects on Earth. As for comets, they are little more than large, dirty snowballs.

Not to be daunted by such trifling and irrelevant details, the logic of cause and effect apparently beyond their wit to comprehend, the primitive inhabitants of our island are quite prepared to accept magic as an explanation.

One of the basic assumptions of astrology is the proposition that people will have different personalities according to their date of birth. Now this may not be quite so far-fetched as it first appears, though any agreement with reality (if agreement exists at all) is more likely to be pure coincidence than anything else. There may well be a seasonal factor which will affect the characters of people to some slight degree depending on what time of the year they were born. Such seasonal variation is not unknown in nature, and in certain species of butterflies, for example, those born in one season will sometimes look quite different from those born in another. But this has nothing to do with the stars except for the one which concerns us most—the sun.

Apart from anything else, an expectant mother's life-style will normally change with the seasons—diet, activity, exposure to sunlight, alcohol intake and so on—and it would be surprising if this did not have some slight effect on the growing child inside her. Another factor involves our sensitivity to the Earth's magnetic field. It has recently been found that most creatures, including bees, human beings, and even some microbes, are sensitive to changes in the Earth's magnetic field (this partly explains the homing ability of pigeons). As some of these changes are seasonal, we have here another possible mechanism for seasonal variation in the characters and personalities of people born around different times of the year. Again the star concerned in these variations is the sun.

There is in fact a small amount of scientific evidence from France, though of a very tenuous and statistical nature, which seems to indicate a very slightly increased chance of certain personality characteristics for people born in certain months of the year. This evidence is regarded as suspect by many scientists but in the light of a rational analysis of the situation it would not be so very surprising if there turned out to be something in it. To say that the seasons produce biological effects, however, is not the same thing as agreeing with the claims normally made for astrology.

One of the craziest of irrational beliefs concerns the spontaneous combustion of human beings, a phenomenon which can supposedly occur at any place and at any time. Remember, we are not discussing the notions of a handful of eccentrics but ideas which have many thousands of believers in Great Britain alone. Although reportedly a very rare occurrence, spontaneous combustion could apparently happen to anybody, including you. No . . . no . . . don't laugh, don't mock the afflicted! As with most of these insane ideas, there is photographic 'evidence' to support it. One of these photographs shows what appears to be a half-cremated corpse lying on a pile of straw in a barn. The straw itself is quite unaffected by the charring, which demonstrates to believers the existence of magical or para-normal forces.

The human body is about 75 percent water (oxide of hydrogen), which has already been burned or combined with oxygen and therefore by definition is a form of ash and no longer combustible. In addition, by preventing the access of free oxygen it is also a great enemy of fire, which is why water is so popular with fire brigades. Most of the remaining constituents of the human body can be burned, but only in special circumstances and with great difficulty unless the water has first been removed.

After the Allied air-raid on the German city of Dresden in 1945, which killed more people than the atomic bomb dropped on Nagasaki, the bodies of the dead being too numerous (about 50,000)† for burial were burned in the city squares on griddles made of railway lines, beneath which burned huge bonfires. Hundreds more had already been reduced to ashes in red-hot air-raid shelters at the heart of the fire-storm. The scale of the conflagration required to produce such burning emphasises the impossibility of spontaneous combustion, let alone in circumstances where straw remains uncharred. Rest assured that if the police were to

†A conservative estimate. Some estimates put the figure at well over 100,000 (the city was packed with refugees at the time).

find the charred remains of your spouse or mother-in-law, or anyone else for that matter, the idea of spontaneous combustion would not enter their calculations for a moment. With the possible exception of belief in witches, who not so long ago also had an affinity with fire, never has a popular superstition been so half-baked both literally and metaphorically. This brings us to the subject of witchcraft, a cultural relic of the Middle Ages but still one of the most widely-held irrational beliefs on our savage island.

The number of people actively taking part in this nonsense is probably many thousands, but pales into insignificance when we consider the millions of ordinary, respectable people who to some extent believe in it. Business is booming in high street occult shops, we are told. An eight week course in witchcraft at a high school in Strathclyde has been cancelled (but only after a public outcry) and the Church of England Commissioners have expressed concern at the prospect of disused churches being bought for use by witches' covens.

The thinking (if that is not too exaggerated a description of the processes involved) and practice of witchcraft is too well known to need further explanation. At least it is one of the more honest forms of delusion in the sense that the word 'magic' and similar terms are openly used, leaving no room for doubt as to the philosophy of believers. Fortune telling by means of tea-cups, palmistry, crystal balls, cards and perhaps even goats' entrails is an allied form of superstition with a large following among the natives, and how many house-wives dare not refuse to buy lucky charms from gipsies for fear of the evil eye?

Many other popular superstitions are taken seriously by people who claim not to believe in witchcraft or magic, yet are still unable to see the logic of cause and effect. It could indeed be bad luck to walk under a ladder if a pot of paint or some other unwelcome missile were to fall on your head. Similarly, it is bad luck to break a mirror as you will probably have to buy a new one, but how throwing salt over your shoulder can avert misfortune or a black cat bring luck except by magic is harder to explain. Belief in ill-omens (broken mirrors, the number 13, especially Friday the 13th, magpies, ravens, etc.) and lucky charms and talismans is widespread.

Considerably less common but probably not rare are people believing in magical beings such as fairies, wee folk, goblins, elves, pixies and so on. Some well known photographs first published in the nineteen twenties and purporting to be of fairies have fooled many people including such respected figures as the author Sir Arthur Conan Doyle.

Another category of supernatural beings widely believed in by the natives includes ghosts, banshees, evil spirits, vampires and the like. To these my earlier remarks regarding souls fairly apply: there is no provision in the DNA 'blue-print' for anything but the physical body. The objection will doubtless be made that these are not physical phenomena. Leaving aside the problems this creates with respect to their ethereal construction and replication, this argument would appear to contradict claims that such beings have seen and photographed. The photons or light particles which stimulate the optic nerve via a photochemical (sight) or produce a chemical reaction on a sensitised film (photograph) are only emitted by atoms and sub-atomic particles—in other words they are physical particles characteristic only of the physical world. It is hard to reconcile this fact with the claim that such beings are made of non-physical material (whatever that might mean).

There is some evidence for the existence of an electric field and a halo of electromagnetic radiation around the human body, and such 'Kirlian auras' have in fact been photographed, but only after artificial electrical stimulation has been applied. This electrical stimulation also produces such auras in other living things and even inanimate objects. As it is stronger in living things, the Kirlian aura has been interpreted by some as 'the life force' (which, as we have seen, does not exist). A more probable explanation is that it is just a weak form of an electrical phenomenon known as St. Elmo's fire, but so weak that a photographic plate is required to detect it. The electrical properties of living things are different from those of dead ones, as any electroencephalograph will confirm, so we should not be unduly surprised at differences in their Kirlian photographs.

Poltergeists, the mischievous spirits which are reputed to fling small household objects about and are suspiciously linked with the presence of mentally disturbed children, make a nice connecting link with our next irrational belief, which is psychokinesis—the idea that physical objects can be moved, bent or broken by the power of thought. It is easy enough to see how this belief has come about, for it is a common delusion amongst the 'natives' that when reality is sufficiently resented and ignored it will obligingly go away, to be replaced by a more acceptable version brought into being by sheer force of will-power and wishful thinking. We will examine some prime examples of this delusion shortly.

Laboratory experiments have shown time and time again that physical effects can only be brought about by the particles and forces of the physical world. It is just as well for the police that this is so, otherwise they might half the time be engaged on a wild goose chase, trying to find human cause for a crime that had been committed by some denizen of the supernatural world! As for the theory that objects can be moved or damaged by the electromagnetic emanations of the brain, the strength of these emanations striking a small object a few feet away would be only the slightest fraction of an erg, totally inadequate to explain the sort of movement or structural deformation often described.

Levitation, the idea that the power of the mind can defy gravity to enable a subject's body to hover in mid-air without visible means of support, is the most extreme and irrational form of this branch of the para-normal. Magnetic levitation is certainly possible with certain metallic or electrically charged objects, but requires electrical power equal but opposite to the force of gravity. The power required to levitate an average ten or eleven-stone human body would therefore be considerable, and though our knowledge of the human brain leaves much to be desired we can say with certainty that it cannot generate power of this magnitude.

The human body can be electrically charged from an external power source so that the hair will rise into the air (stand on end), but will not hold sufficient charge to levitate itself, and any attempt to make it do so would not only fail but be extremely dangerous. Various tricks are possible with electro-magnets and other forms of deception, but this is not levitation in the sense which concerns us here. Some well known photographs of 'levitation' have been exposed as fakes by disillusioned ex-mystics, and were simply 'spot-the-ball' type pictures where the action was frozen by the camera, leaving the subject apparently floating in mid-air.

Acoustic levitation is also possible in certain cases, but again the principle is the same: a source of power (in this case sound) is equal but opposed to the force of gravity. The sound produced by the human body might conceivably levitate a pollen grain but nothing significantly more substantial. As for wishful thinking, it will no more levitate the human body than it will achieve any of the other marvels which hopeful devotees are constantly attempting.

On the subject of anti-gravity, but with no bearing on human levitation, it is not known with any certainty whether the gravitational force of anti-matter is precisely identical to that of matter. The proposition that matter and anti-matter might repel each other gravitationally is generally discounted, but is not testable in the laboratory because the gravitational interaction is undetectable at the atomic or sub-atomic level at which anti-matter is available on Earth.

Another form of the para-normal with a wide following amongst the gullible is 'pyramid power'. The suggestion is that this shape has mystical power to concentrate the 'energy of the universe' (whatever that is) towards some useful end. Believers claim that enhanced mental powers and other para-normal effects such as the mysterious self-sharpening of razor blades can be obtained by placing oneself or the relevant object inside a pyramid. The material from which the pyramid is constructed, whether wood, stone, metal or whatever, is apparently irrelevant.

Pyramid power generated sufficient public interest a few years ago to prompt some serious scientific tests, but results, predictably, were negative. Having said that, however, there may well be a psychological effect among ardent believers analogous to the well known placebo effect in medicine—make-believe has a curious tendency to be self-fulfilling, as we shall see (though in a strictly logical and unmysterious way). As for real effects, pyramids are no more effective than domes, cones, cubes or any other shapes. If you are still not convinced, this could be tested experimentally for yourself without too much difficulty, using blunt razor blades to test the effects of various shapes. The magical properties of certain geometrical designs is a concept older than recorded history, and it is in these ideas rather than in science that pyramid power has its origins.

The one aspect of the so-called para-normal which cannot easily be dismissed is telepathy (mind reading), which if it exists at all might also help explain certain other alleged phenomena sometimes referred to as extra-sensory perception. The human brain is a very complex instrument, still not well understood, and is the source of infinitely subtle electric currents and electromagnetic radiations. Because these currents and emanations are so very weak the chances of telepathy ultimately proving to be a valid proposition are slender, but at present it seems just possible that some people are able to receive and interpret feeble electromagnetic signals arising from human thought processes.

Surprisingly faint signals can be picked up over long distances even by clumsy man-made instruments. We should not forget that sensitivity to the Earth's magnetic field amounting almost to a sixth sense has only recently been detected in humans and other animals, and may be viewed as a form of extra-sensory perception. Whether telepathy will prove to be a similar case is anybody's guess. The trouble is that several people who might possibly have this power to a limited degree are showmen, and embellish their para-normal abilities (if any) with trickery. It is therefore difficult to sort the gold from the dross, or even to be sure if there is any gold at all.

More exotic para-normal powers are, however, another matter entirely. These, if they existed, could be (and therefore would be) used as weapons of war. For example, the ability to bend the compass and instrument needles of enemy aircraft by psychokinesis would have obvious military applications.

Throughout history there have been numerous attempts to use magic as a weapon of war, and all instances which are well authenticated have been conspicuously unsuccessful. When fighting Hereward the Wake in the fens, William the Conqueror is said to have installed a witch in a wooden tower to cast spells on the enemy. The wily Saxon rebel promptly set fire to the rushes round about, and the 'wicked witch' in her wooden tower was burned to death. Just what went wrong with her 'magic' has not been explained from that day to this! In the mid 1960s the Simba guerrillas, on a rampage of murder and atrocities in the Congo, were told by their witch-doctors that the bullets of the enemy would turn to water. To nobody's great surprise but their own this failed to happen, with predictable results which bore a remarkable resemblance to poetic justice.

So magic has been discredited at every turn, both as an explanation of the universe and as a weapon of war. Spiritualism and communication with the dead are also quite impossible for reasons already explained, and would only be possible if the laws of physics and biology were abolished. In so far as most natives believe in one or more of the primitive superstitions listed above (plus several others which have escaped mention) it will be seen that most of them still believe in some form of magic! People claiming to have the sort of powers which (if genuine) would soon make them millionaires are nevertheless not millionaires, a very suspicious circumstance which is paralleled by those gentlemen who offer to sell you the means to make a fortune on the football pools. One wonders why their philanthropy forbids them to use their powers on their own behalf.

We will now examine those irrational beliefs which do not involve magic in the usual sense of the word. These non-magical irrationalities may be further sub-divided into those relating to normal, everyday events and those relating to unusual phenomena.

Two of the most preposterous and widespread of everyday irrational beliefs are, firstly, the idea that reality is unimportant and there is no real need to look at it at all and, secondly, that if one must be so boorish as to attempt to explain or describe it, this can be done without the necessity for anything so mundane or indelicate as calling a spade a spade. A little honesty is all it takes to see that this philosophy is about as appropriate to the real world as it would be to a book on gardening, let alone a book on reality!

The popular attitude to books and the standard of honesty among authors and journalists is particularly instructive to students of reality. The overwhelming demand for fiction in preference to non-fiction is significant and entirely consistent with the idea that most people are living in a dream world. You have only to go to your local library to see crowds milling around the fiction shelves, while the non-fiction and science shelves are neglected by comparison. Much of this fiction is mind-destroying fantasy fodder which seems to be directed at one sex in particular.

The gross distortion of fact is a feature of popular literature and authors seem to believe that they cannot sell books without it. Westerns, for example, bear very little resemblance to the Old West as it really was. History, especially recent history, is a veritable treasure trove of delusion and self-deception masquerading as fact. While we should not forget Liddell Hart's dictum that the word 'official' often cancels the word 'history', historians themselves are probably less to blame than the popular imagination, the need people have to believe what they find congenial.

Get an authoritative book on the strength and equipment of the opposing armies in France in 1940 or the subsequent Dunkirk evacuation and Battle of Britain by a thorough and impartial historian (preferably not an official historian) and it will be very surprising if you are not relieved of a few misconceptions and left lying amid the wreckage of your shattered illusions. The proportion of German divisions deployed against the British in 1944 (about ten percent) will provide another unwelcome surprise for those who would like to believe we won the war almost singlehanded, while a history of slavery (still practised in a dozen or so countries to this very day) will come as a shock to those who thought it was all the fault of the Europeans. Yes, the better historians, of the type approved by Liddell Hart, may give us a rude awakening when they can break through our self-imposed censorship.

Another example of popular mythology is the idea that a man can be conveniently anaesthetised by a blow on the head and probably several people die every year in Great Britain alone as a consequence of this particular delusion. Private detectives readily admit that their job bears little resemblance to that portrayed in fiction, and in the world of espionage (human nature being what it is) fact sometimes imitates fiction rather than the reverse.

There is generally hostile attitude to the philosophy of total reality among the islanders, and many do not bother to disguise their antipathy when the subject is broached. Their opposition is characterised by a conspicuous lack of reasoned argument and an irrational head-in-the-sand defence against any threat to their illusions or sacred cows. There is also a tendency to blame total realists or bearers of bad tidings (the two are not necessarily synonymous) for stating unwelcome facts, forgetting that total realists merely state the facts and are not personally responsible for them. Reality has become the new unmentionable subject, and because publishers are subject to the same human failings as most of their readers it is extremely difficult to get them even to consider a manuscript which deals frankly with it. The result is a dearth of such books in the market place which verges on the sinister.

A number of human idiosyncracies which seem harmless enough at first sight are rather less innocent when we examine them more closely. The popularity of euphemisms, for example, serves to blur the boundaries of reality when that reality is harsh, distasteful or unpleasant: hence 'put to sleep', meaning to kill or to carry out euthanasia. 'Gay', meaning homosexual, is a particularly sad perversion of the English language and has destroyed one of our most attractive words. Then we have 'massacre', a term sometimes applied to victory by an enemy but never to victory by one's own forces. 'Freedom fighter'— someone as often as not fighting for the imposition of a Marxist one-party state and prepared to deny to innocent civilians, who only want to be left alone, freedoms far more important and fundamental than the ones he is fighting for. Euphemisms are particularly beloved of bureaucrats and totalitarian states, the notorious 'final solution' being perhaps the best known example. Certain governments do not shoot down heavily laden civilian airliners which stray into their air-space, they 'stop the flight'. Bombs dropped in the wrong place are 'incontinent ordnance'. The reader will no doubt think of other examples for himself. Non-realists can seldom bring themselves to call a spade a spade and are inclined to take umbrage when others do so.

The popularity of ritual and symbolism in our society also seems innocuous, but deserves closer inspection. There may be a place for ritual in state ceremony to lend dignity and spectacle to some solemn occasion (Dr. Göebbels was convinced of it), but blind obedience to a stylised and mechanical set of procedures is often used to obviate the need for an individual to think for himself, especially when the authorities doubt his ability to do so to their liking. For this reason it is often used in the Army, where it is called 'drill', and its purpose is to transform men into robots—a perfectly legitimate practice, as a rule, in the interests of military efficiency. In the animal kingdom too this role of ritual and symbolism as a substitute for intellect is prominent, for animals have only a limited capacity to think for themselves. So ritual behaviour has been with us since the Permian at least, and can thus be traced back to some of our most distant ancestors.

Symbolism has other legitimate uses, for example the symbols of writing or mathematics, or the insignia of military aircraft. The trouble comes when we use symbols not as indispensable and necessarily stylised representations of reality but as substitutes for it. It is much easier for us to acquire a status symbol than to actually improve ourselves, to flaunt a virility symbol than to truly be man, or to display the appropriate symbol rather than cultivate some desirable aspect of mind or body which requires hard work. More foolish than the natives who actually employ such symbols are those prepared to accept them as substitutes for the reality they are supposed to represent.

Mysticism has always held an important place in primitive societies, but its popularity in a modern industrial state on the brink of the twenty-first century may puzzle some people and calls for an explanation. Again seemingly harmless ideas turn out to be not so innocent.

In order to allow make-believe to masquerade as reality a degree of fog and obscurity is required, but this makes a very odd vehicle for the presentation of truth. The purpose of mysticism (though only a reformed mystic would be prepared to admit it) has always been to lend plausibility to basically irrational ideas and to project the illusion of knowledge rather than its substance. In prehistoric or even Biblical times this was almost excusable, as man's explorations into the nature of reality had hardly begun, and relevant knowledge was almost non-existent. Today the purpose of mysticism is the same: to give the illusion of understanding without the painful acquisition of facts and the harsh exercise of logic which must accompany any clear perception of reality.

In common with astrology, fortune telling and other imaginary phenomena of the para-normal, there is an underlying dishonesty which, under a smoke-screen of ambiguity and self-deception, seeks to blur reality and prevent understanding rather than promote it. As with other forms of 'virtuous' dishonesty (i.e. dishonesty which is accepted or even required by society) practitioners do not necessarily *feel* themselves to be dishonest.

It is sometimes said that scientific jargon, which tends to be incomprehensible to the layman, is a form of mysticism. There is, however, an important difference. Scientific jargon is simply a form of shorthand which, far from obscuring the truth, makes for clarity and conciseness when used by the initiated. Most scientists can put their ideas into everyday language comprehensible to the man in the street and still aim for clarity, though partly at the expense of brevity and conciseness. But whatever the scientists' command of language there is a limit to the extent that complex issues can be simplified.

The censorship, whether imposed or voluntary, which is the friend of ignorance and delusion is the enemy of science, which finds it abhorrent. This should speak volumes to those who love truth. No wonder scientists avoid mysticism—their purpose is to clarify rather than conceal, to appeal to logic and reason rather than ignorance and superstition. A recent book on mysticism refers to meditation, and describes it as similar to sleep! This explains a lot about the philosophy and beliefs of converts, most of whom, despite their protestations to the contrary, are looking for tranquility and peace of mind rather than truth. The bloody battlefields of the war against ignorance and superstition are hardly the place for seekers of tranquility. In the quest for truth another sort of meditation is necessary: thinking for yourself, which is often hard work and never at all like sleep. Not surprisingly, many mystics favour the use of hallucinatory drugs to produce the tranquility and 'insight' they crave.

The widespread use of 'psychedelics' (the concept of which is based on drug-induced hallucination) gives a much-sought-after air of unreality at discos and is not without disturbing psychological implications, but far more serious is the ever-growing use of drugs, both legal (tranquilisers, 'magic mushrooms') and illegal (narcotics, L.S.D., etc.), to blur the boundaries of reality when they intrude into our dream world. Forty five million prescriptions for tranquilisers were issued in Britain in 1984. In the same year nearly a billion pounds was spent by the N.H.S. on tranquilisers and related drugs, mainly for use as a substitute for moral fibre in people wishing to avoid reality, while others died for lack of funds to provide them with renal dialysis. So a sizeable percentage of the population are not only dreaming their own lives away but, more unfairly, helping to dream away someone else's as well.

The traffic in illegal drugs is also a multi-billion pound industry which causes many deaths every year. Far from having levelled out, consumption of tranquilisers and other drugs is rising steeply. Will we see the day when women's magazines print recipes for Valium soup or Librium on toast, or perhaps Barbiturates à l'anglais? But drugs are only an additional aid to creating that cosy aura of make-believe with which most of us like to surround ourselves. This more natural cloud-cuckoo-land, though safer than the drug induced variety, is nothing like as harmless as is generally believed. The cosiness is a mirage, and the real world becomes a little harder to face each time we evade it.

A reality-hating society of fantasising make-believe addicts is naturally predisposed toward the taking of hallucinatory drugs, and it is this conditioned predisposition, I believe, which lies at the heart of the problem. Our next example must surely be one of the most monumental confidence tricks of all time, and has successfully hoaxed mankind from prehistoric times right down to the present day.

'Class' in its conventional sense—the notion that one man is superior or inferior to another by inherited social accident of birth: royalty, aristocracy,'untouchables' (an absurd invention of the Hindu caste system which still has credence with some immigrant inhabitants of our savage island) and so on—is a figment of the human imagination produced by an interaction of ego, ignorance, and self-deception. It has no substance in reality, as it is not embodied in the nucleotide sequence of the DNA molecule, the genetic code by which *all* inherited characteristics are passed on. There is no gene for noble blood, whatever that might be.

The reason this colossal hoax is still so widely accepted and perpetuated in the so-called 'Age of Reason' is probably that many people have a vested interest in so doing. Obviously we are not all genetically equal at birth, but class in the sense we have been discussing exists solely in people's imaginations, and is only real in the sense that the Emperor's new clothes were real. Superior quality of an inherited nature is just as likely to be found in the offspring of serfs and untouchables as in royalty. This is not a political statement but a scientific fact (the merits and demerits of monarchy as a system of government are quite a separate argument, and beyond the scope of this book).

Official recognition of the magical healing powers of kings (the ability to heal diseases by touch) was finally abandoned in England during the eighteenth century, though belief lingered on in the public mind. In France the idea was not officially abandoned until well into the nineteenth century. The association of royalty with fairy stories is well known and has a long tradition. It was perhaps this mystical aspect that the former Lord Altrincham had in mind when he said that we are in danger of making royalty into a religion.

It will have been seen from the contents of previous chapters that the attributes and qualities of any organism (and hence class in the true sense) are largely genetic in origin and therefore incorporated into the structure of the nucleotide sequence. In biologists' slang this hereditary factor is known as 'nature', as opposed to the influence of external factors such as feeding or environment, which are regarded as 'nurture'. Looks, physique, intelligence and character, although dictated by the genetic code, will be powerfully influenced by such external factors as mental stimulation, home background or the amount of protein in the diet. This accounts for the apparently superior quality of some members of the aristocracy, whose supposed 'good breeding' is actually a result of nurture rather than nature.

Obviously, any factor which cannot be encoded in the DNA molecule to be later enhanced or suppressed by such external influences cannot be heritable, and will therefore produce no increased mental or physical performance in succeeding generations. This applies to any organism, whether dog, falcon, horse, man, or whatever. Even the inheritance of desirable arrangements of genes is a very chancy business, owing to the inherently random nature of genetic processes.

This fact was recognised by the famous playwright George Bernard Shaw when a beautiful actress of his acquaintance proposed marriage. "With my beauty and your brains," she said, "think what splendid children we could have." "Ah yes," replied Shaw, "but suppose they have my looks and your brains!" This makes the serious point that even parents of genuine high quality will not necessarily pass this quality on to their offspring—a fact all too well known to breeders of race-horses and other animals. The reverse is also true of course, as anyone who has seen the high class progeny which occasionally spring from nondescript parents will testify.

Genetic defects, on the other hand, are just as likely to occur in royalty as in anyone else, perhaps more so because of inbreeding. Queen Victoria, for example, was a carrier of the genetic defect haemophilia which, although it died out in the British royal line, was passed on to two other European royal families through her descendants. Haemophilia inhibits clotting of the blood, so sufferers risk bleeding to death. The condition is carried by females, who are otherwise unaffected, and manifests itself only in male descendants.

But enough of royalty: what about the workers? Another myth, which is probably less widespread among our own natives than among the inhabitants of Communist countries, is that these countries are a workers' paradise. Why walls, fences and mine-fields, and armed guards with orders to shoot first and ask questions afterwards, should be necessary to keep workers inside their 'paradise' (and these measures are only partly successful) is not explained by enthusiasts. The idea that these border defences are necessary to keep out hordes of illegal immigrants eager to sample the delights of a workers' paradise does not stand up to examination, and I doubt if there is anyone who truly believes it. A popular explanation in East Germany is that such defences are necessary ". . . to keep out Nazis"!

Most convicted criminals claim to be innocent, but are more likely to be lying than out of touch with reality, at least on this particular issue. However, with regard to those who take their claims seriously (mainly their families and particularly their wives and mothers—ironically enough the very people who should know them best), it is another matter. Are we seriously to believe that our prisons are full of innocent men, that the police and judiciary are engaged in a nation-wide conspiracy to pervert the course of justice? Here we have a splendid example of wishful thinking substituted for reality. Time and again we find our primitive natives seeing what they want to see rather than what is actually there. Indeed, so common is this trait that it is accepted as normal! This tendency for people to believe what they find agreeable and reject the rest has been confirmed by scientists in a number of separate experiments using a variety of psychological tests, and every time the results were unequivocal. There are wrongful convictions, it is true, but these are very rare and amount to only a fraction of one percent of those claiming to be innocent. British law, though not infallible and leaving much to be desired, the best in the world by default rather than merit, does stop short of sinister conspiracies to convict thousands of innocent men.

Another common irrationality is the belief that people improve themselves by means of acquiring money, and that rich people are somehow 'better' than those less well off. Surely no one would suggest that a mediocre painting could be improved by the addition of sufficient money so that it could be hung in the Royal Academy, or that a sheepdog could be improved in this way in order to win the National Sheepdog Trials? Surely a cat that had been left a fortune would not be accepted at a cat show as being for this reason better than other cats? Many people will openly admit that they not only equate self improvement with the possession of material wealth, but material wealth with civilisation, and claim that one is a measure of the other. I suggest that awareness of reality would be a far better yardstick, and it is no accident that these people tend to be among the least civilised except, perhaps, by their own absurd definition.

We have only to remember President Idi Amin of Uganda in his palace at Kampala, and compare him to the young Abraham Lincoln doing his schoolwork with charcoal on the back of a shovel in the humble log cabin which was his home, to see that civilisation is independent of personal wealth and more to be equated with riches of the mind. For all his finery, Idi Amin and others like him will never achieve a shadow of the civilisation of the boy who could not afford an exercise book.† (The infamous and now deposed Emperor Bokassa of the Central African Rebublic, notorious both for the lavishness of his coronation and the repression and exploitation of his subjects, is another example that springs to mind.) The comparison is only one of many that could be drawn to show that material possessions are not a yardstick of civilisation but a spin-off or by-product from it. Einstein or Lincoln in a hovel would have been more civilised that most other men in a palace.

The facts are that mental or physical improvements to mice or men cannot be brought about by the mere acquisition of wealth, though money may sometimes buy the *means* to effect improvements (or for that matter further degeneration). A rich idiot does not become a genius but remains an idiot plus money.

Even in specialist scientific subjects such as the study of birds of prey (a special interest of mine) people commonly see what they expect to see regardless of what is actually there, and believe in such mythical creatures as 275 mph., nest-building peregrines, griffon vultures which eat 6 kg. in a single meal, and a hundred and one similar absurd fictions which have no substance in reality and often defy the laws of arithmetic. A book could be written on this topic alone. Of course, the peregrine may sometimes *look* as though it could do nearly 300 mph and the griffon tucking into a carcase may seem to have every intention of consuming 6 kg. (almost its own weight) but the true figures in both cases are only a third of those quoted. As with other falcons, the peregrine builds no nest.

The authors of these and many other old wives' tales are in most cases respected ornithologists, whose ideas of rare and threatened species are also most peculiar. The public are under the quaint delusion that peregrines, ospreys, golden eagles and the like are rare species in danger of extinction, while the experts themselves sometimes seem never to have heard of the real rarities. Kestrels and black kites, listed as endangered species, are really only in danger of exceeding world populations of ten million, but have you ever heard of Eleanora's falcon, a European bird of prey numbering only about 3,000 pairs and less than one percent of total world numbers of peregrines? Most of the popular 'endangered species' are in fact fairly common and likely to remain so, though in some cases there may be a retraction of range. Only about half-a-dozen are in any real danger of extinction within the next hundred years, and comprise species which the majority of the public have never even heard of. The

†I have assumed that readers are familiar with the make-believe, fantasy aspects of Amin but, if not, his 'decorations' (V.C., self-awarded) and 'titles' (Conqueror of the British Empire) give some idea of the character of the man.

term 'endangered' has become almost meaningless, and it is about time some precise scientific definition was drawn up. A word appropriate to the Mauritius kestrel or the California condor is often applied to birds which exceed them in number more than ten thousand times over.

This make-believe world is by no means confined to the study of birds of prey. As early as 1978 so-called experts were predicting total destruction of the world's rain-forests within twenty years and I was contradicting them. Analysis by the principles of total reality indicates that less than ten percent will have gone within this time, mostly in low lying areas which are readily accessible. So you have less than fifteen years to wait to see who was right—me or the 'experts'. In fact the answer should be apparent long before then. "But isn't it better to be on the safe side?" you may say. Yes, but it is better still to be accurate and in close touch with reality.

"Got you!" non-realists will be saying triumphantly. "Haven't you been extolling the virtues of science? And now you tell us that the experts cannot be relied on!" No. I have said that some experts cannot be relied on. Scientists are not a special breed of men but for the most part ordinary (one might almost say normal) irrational human beings who have to wear the strait-jacket of logic imposed on them by science. Some wear this strait-jacket with better grace than others. A few will cast it aside given half a chance, to make lunatic excursions into the land of make-believe which is unfortunately peopled by the majority of the human race.

There have been several notorious cases of bogus experiments and falsification of data, but these were usually swiftly detected by the scientific community. Those misguided enough to chance their arms do so mainly in popular works and are more restrained when writing scientific papers for learned societies and periodicals. Just why the public should be less entitled to the truth remains one of nature's unexplained mysteries.

The willingness of a small minority of scientists to see what they want to see is well documented, and several cases of collective self-deception involving no intentional dishonesty are known. There is little doubt that other less clear cut examples go unrecorded. This does not mean, however, that most or even a significant proportion of scientific investigation is unreliable.

Scientists usually indulge themselves in flights of fancy and self-deception only when they think they can get away with it, so such practices tend to occur in scientific backwaters and tributaries where competition is less fierce and they are less closely observed. This is in contrast to mainstream, big money science such as physics, molecular biology, or astronomy where, apart from anything else, departure from reality or failure to reach the highest standards of accuracy can have consequences almost too frightful to contemplate. You cannot afford to be several hundred percent out when calculating the critical mass of a nuclear reactor, de-activating a virus, or calculating the escape velocity of a lunar excursion module.

One of the things which hold them in check is a fear of ridicule, particularly when they have a reputation to consider. They know full well that many of their colleagues would be only too glad to bring them down by pouncing on the slightest mistake or self-indulgence. With many scientists however, honesty, precision and professional integrity become a way of life; they acquire a reverence for truth and live only for their work. Such people are often relatively uninterested in the material things by which the rest of us set so much store.

Scientists are particularly careful to avoid wishful thinking, which they do mainly by a system of 'controls' and 'double blinds'—dummy experiments, the results of which are later compared to those of the real thing. It is important for experiments to be repeatable, and results are seldom accepted as valid until they have been repeated by an independent authority.

In science no one is beyond question or criticism, and even the work of great men like Darwin, Pasteur and Einstein is frequently subject to test and verification by people hoping to prove them wrong. For all these reasons we may have confidence that science is built upon a rock, and though storms may sometimes blow off a shoddy tile (or perhaps even bring down a chimney) the main edifice will stand firm. Having thus allayed your fears or crushed your hopes of finding an escape route, whichever may be the case, we shall now continue our survey of common irrational beliefs.

It is often hard to say where normal aversion to reality ends and madness, dishonesty or just plain wickedness begins. Those who have gone out of their way to commit senseless and unauthorised atrocities (as in Vietnam, for instance, where such behaviour amounted to an American-sponsored recruiting campaign for the Viet Cong) will then take part in demonstrations against the brutality of war. A whole catalogue of crimes, ranging from mere censorship to torture, murder and genocide, will be carried out by people who claim they are doing it for God! People whose families befoul and vandalise their recently built housing estates will quote the resulting stink and shabbiness to support their claim for rehousing. Others will go on a rampage of crime and arson to prove that they are honest, law abiding citizens who have been unfairly picked on by the police.

Hostages who only a short while before were being threatened with death and had seen fellow hostages executed will become convinced that their captors are saintly and heroic freedom fighters, God's gift to humanity. This phenomenon is well know to police psychologists and hostage negotiators, who even have a name for it—the Stockholm effect—and can cite several extreme examples. It seems that many people have inherited from our primitive ancestors the desire to grovel and abase themselves before someone in authority, even when this is no longer necessitated by the demands of expediency. Dog-like, they revere those whose foot is on their necks, perhaps from a mistaken sense of gratitude that their oppressors did not deliver the coup de grace.

In such a situation it is human nature, with some people at least, to bite the hand that feeds them and lick the boot that kicks them. The phenomenon is put to good use by police and military intelligence gatherers, who alternate 'kindly' with brutal interrogators to put suspects under psychological pressure.

It is said that there is no one we hate so much as someone to whom we have done an injustice. How very odd, when logic so clearly indicates that the hateful person is none other than ourselves! With many people, however, this logical and more accurate conclusion is unacceptable on emotional grounds. The trouble is that most of us are too soft hearted. When painful self-discipline is required to effect necessary improvements to our flabby minds or bodies we cannot bear to be harsh, but continue to lavish kindness and affection on ourselves. "Eat yourself fitter!" says the advertisement, knowing full well this weakness of human nature. The truth is more accurately summed up by the slogan of the keep fit enthusiasts: "No gain without pain!"

Behaviour regarded as criminal when committed by our enemies will be seen as reasonable and justified when committed by ourselves or by people or governments of whom we approve. 'Freedom fighters' will go to almost any lengths and make almost any sacrifice to install a tyranny as bad as or worse than the one they wish to overthrow. Occupying armies will ignore the oft-repeated lessons of history and foolishly continue to make life harder for themselves by launching massive recruiting campaigns on behalf of the enemy resistance. Will they never learn? Moral molehills will be magnified to the proportions of moral mountains to preoccupy the attention of politicians while the real moral mountains go apparently almost unnoticed. But then, why should the views of the majority on moral matters be any more rational and in touch with reality than their views on anything else? Anti-racists will remain blissfully unaware of their own inverted racism which judges infringements of human rights more on the skin colour of the perpetrators than the enormity of the crimes.

One of the commonest irrationalities of all is the belief that those who cultivate an obsession with trivia and scarcely bother to hide their loathing of logic, fact and reason, practising a self-imposed censorship as ruthless as any in Pravda (but as seemingly benign as certain British 'newspapers' which are equally devoid of news), are nevertheless just as in touch with reality as anyone else. Intuition and wishful thinking are seen as much more reliable guides to truth than the mundane acquisition of data and the harsh exercise of logic. People who abhor the excesses of various over-human dictators, tyrants and terrorists will make not the slightest effort to stamp out the lesser but similar manifestations of human nature to be found within themselves. They are in favour of the infection, it seems, but not the disease to which it leads! The main difference between the tyrants who are continually making headlines, and the sort of man you pass every day in the street, is that the man in the street lacks opportunity. This is why power so often corrupts and absolute power corrupts absolutely.

Leaving aside the popular dislike of reality which is so far not recognised as a phobia by psychologists, the vast majority of people (perhaps as many as 90 percent) have an irrational fear of some harmless object or creature. Sparrows, cats, spiders, bats and open spaces are among the commoner ones.

Contrary to experience and common-sense, many people (including some trade union leaders) maintain that excessive pay settlements have nothing to do with inflation. Other than to say that this is not the opinion of any qualified economics expert I do not propose to go more fully into the question here, but it is interesting to note that militant trade unionism, like militant religion, can sometimes produce the very evils it was designed to combat—a paradox seldom taken into account when considering the morality of not joining a trade union.

There are also those who appear to think that wages restraint, like awareness of reality, is fine so long as it only applies to other people, others who think that strikes are always justified in Poland but never in England. Most dangerous of all are people who think they can bring the country crashing down in ruins about them and emerge unscathed from the wreckage. The only people who might emerge from such a disaster better off than before are our future commissars or gauleiters, and then only if they are lucky.

Whilst on the subject of politics, we have all seen examples of opposition parties claiming that high inflation and unemployment are mainly due to the ineptitude of their opponents and suggesting that things would have been very different under their rule. Not only does the arithmetic of their calculations often look contradictory to the logical eye, but what do we find when these people take office? Not that they have the magic panacea for our economic ills as they had previously claimed, but that even years later these problems are substantially worse than before. The root of the trouble is that they insist on clinging to doctrinaire ideas that seemed alright in opposition but have very little bearing on reality. They have also sold their souls to certain of their supporters with promises which sometimes bear a suspicious resemblance to bribery.

"But everyone knows politicians are dishonest!" you may say. Not many politicians would agree with you, so if we accept their assurances that they are not knaves we are left with the conclusion that they must be fools. What is beyond dispute is that believers in almost all the other nonsenses I describe are to be found in their ranks. By the very nature of politics, politicians tend to be actors rather than scientists, and their talent more for manipulating people than for rational thought.

A number of the more primitive savages believe that a campaign of murder and atrocities will get the British Army out of part of Britain when in fact nothing could be more calculated to make it stay—yet another example of militant make-believe having the opposite effect from the one intended. An escalation of violence would be far more likely to result in an increased Army presence than a reduction, but those who compare the British to Nazis or the Gestapo (a favourite analogy of terrorists) are either deluding themselves or extremely ignorant of what went on in Hitler's Third Reich.

The claims of terrorists, wherever they operate, are to some extent self-fulfilling. They must of necessity provoke a response which involves some curtailment of civil liberties and a superficial resemblance to the beginnings of a police state. The superficial and highly censored view of reality adopted by terrorists (who are *not* as a rule synonymous with freedom fighters) accounts for their willingness to deny to others the freedoms they so stridently claim for themselves, and the seemingly Alice-in-Wonderland abandon with which they plant bombs in public places.

The truth is that the tyranny they decry is a manifestation of the same unreasonableness and irrationality with which they themselves have been programmed. The opposing doctrines of extreme left and right are but different sides of the same coin. Their propensity to ignore the democratic process, where this is available, is no doubt because they find it too slow for their liking and, worse still, likely to favour the majority. (The majority are a pretty sickening lot, I agree, but so too are undemocratic and reality-hating minorities.)

The situation is further complicated by the fact that in some countries government forces act as terrorists themselves, and in a few cases the insurgents are actually outdone in barbarity and mindlessness by the supposed forces of law and order. But I digress. Let us return to perfidious Albion, which is the country that concerns us most.

Many of the natives, perhaps the majority, find the make-believe world of popular soap operas more interesting than reality, and the 'shooting' of one fictional character was headline news for weeks, as were the imaginary marital problems of another. A small but disconcertingly large number appear to think that the fictional characters are real people and the dramas portrayed are real-life situations. It may be normal to blur the boundary between fantasy and reality, and it may be that these people are only hoaxers or practical jokers, but there is evidence to suggest that some of them have abandoned reality to an alarming degree. A favourite theme of these 'soap operas' seems to be that nice guys finish last and the rich and the powerful are just as much animals as the less exalted—a theme which many of the less exalted actually find flattering! The anti-hero has become so admired that we are in danger of becoming a nation of anti-heroes.

Many other examples of irrationalities are provided by everyday life and attitudes—especially by the activities of bureaucrats, to the eternal delight and profit of newspaper reporters and television watchdog and consumer programmes. To be fair, there must be many bureaucrats doing an intelligent and competent job, but the sort of irrationality, inefficiency and aversion to reality mentioned above is commonplace and not confined to isolated incidents, as many of you will know to your cost. The investigation of almost any water authority would provide further evidence if it were needed, and such investigations have in fact been carried out. So common are its manifestations that one might be excused for thinking that persons having this bureaucratic mentality are selected by the Civil Service rather than weeded out. Space considerations forbid detailed examples of a subject which deserves a book to itself.

Nazi Germany was in some ways a magnificent example of bureaucracy gone mad. In case you think the analogy too fanciful, it should be explained that one of the striking features of most war criminals, especially at the lower end of the scale, is their ordinariness. They would not normally stand out in a crowd. Contrary to what you might think they are not solitary individuals who amuse themselves by plucking the wings off flies and setting fire to cats, or loners and social misfits who brutalise their wives and lead a furtive, twilight existence of crime and sadism.

The unpalatable truth is that most of them are just like you or the chap next door. Heinrich Himmler (head of the S.S. and Gestapo) was, from the point of view of friends and neighbours, a charming family man who has been likened many times to a bespectacled, mild mannered Civil Service clerk, and on the few occasions when he witnessed an execution he was visibly upset. Klaus Barbie (the infamous 'Butcher of Lyons' now awaiting trial) is described by his daughter as a kind and loving father, a delightful man who was only doing his job. It cannot be emphasised too strongly that war criminals are usually just ordinary people, especially in their desire for conformity. Owing to censorship, which is largely self imposed, they have a highly distorted view of reality and are usually not deep thinkers. Neither are they confined to any particular culture or nationality, though they may be more common in some cultures than in others.

A whole army of administrators and participants at every sort of level was required to organise the arrests, deportations, and the machinery of the Nazi death camps, from the clerk who typed out the letter requesting supplies of cyanide for the gas chambers to the civilian contractors who built the notorious crematoria ovens of Auschwitz. Would you have typed that letter? Would you have submitted a tender for those ovens? The chances are that you would. The S.S., by the way, was primarily a political rather than a military organisation, and numbered at least one bishop amongst its ranks.

Much of the work inside the camps was done by the Jews themselves under duress, but this does not alter the fact that many thousands of non-Jews were involved directly or indirectly in the extermination programme. There were several instances where Germans refused to comply with orders which they rightly regarded as dishonourable and inhuman, and contrary to popular belief they were not severely punished. It *was* possible to disobey.

Some interesting and well publicised experiments were performed in America a few years ago which demonstrate the potential for evil in almost every one of us. Psychologists recruited volunteers selected at random under the pretext that they were required ". . . to assist in a scientific experiment." Their task was to operate a lever which administered a series of electric shocks, graded from mild to severe, to the supposed subject of the experiment, an elderly man fitted with electrodes and strapped into a chair.

When the subject failed to produce the right response to the psychologist's questions, these assistants were asked to administer electric shocks as a punishment. The first wrong answer was punished by a mild shock which produced only a slight twitch. The next offence was punished by a more severe shock which elicited signs of pain from the subject. If he persisted in giving wrong answers the volunteers were told to increase the voltage, which caused the subject to strain against his bonds, uttering cries of pain. The experimenters usually went on to use a stronger current which made the subject jerk in his chair, shrieking in agony and begging his tormentors to stop, but when he still failed to give correct responses to the interrogator's questions they were told to administer punishment with an even higher voltage.

What the volunteers did not know was that the man in the chair was only an actor and that they themselves were the true subjects of the experiment! A very small percentage of the volunteers, having given the first shock (there was, of course, no actual current), refused to go any further and opted out of the experiment, but the vast majority went almost to the limit and a small but substantial number did everything they were asked with evident signs of enthusiasm. The chances that you would have drawn the line after administering only the first mild shock, or even the slightly stronger one, are very, very slender.

It is a chastening thought that concentration camps were first invented by the British during the Boer War (1899–1902), some thirty-five years before Hitler followed suit but still within living memory. Although lacking the later refinement of gas chambers, these were not concentration camps in name only but Nazi-style death camps in which 30,000 people died, mostly Afrikaner women and children. These facts, which are supported by damning photographic evidence, are not nearly so well known as they deserve to be. Thirty thousand people may not sound very many, but to the tiny Boer community (fewer than the population of an average English county) the death toll took the proportions of a holocaust. This is not the only skeleton in our military cupboard, though the others (apart from Dresden) are comparatively minor.

Most people are chameleon characters who are slaves to conformity within their group, however unconventional the group itself may appear. They have hardly any character of their own but adopt one which suits their surroundings and are neither fundamentally good nor evil, honest nor dishonest, but a reflection of the minority of stronger characters around them. In a situation where the small percentage of brave and courageous men predominate, the chameleon characters will reflect the prevailing mood and themselves become brave. If a small percentage of cowards set the mood, the herd will tend to follow. When a small percentage of truly vicious and evil men predominate, they can rely on the chameleons to conform, as history has shown many times.

The majority of people taking part in horrific atrocities were not psychopaths who had spent their lives doing whatever it is that psychopaths do, but ordinary people who (as they saw it) were just doing their job. Most of them were pleasant and charming family men in their private lives, the sort of people that no one would mind having as neighbours or drinking companions. Do you really want to be like them? From now on perhaps the term 'little Hitlers' will take on a new depth of meaning for you. The balance between good and evil in human nature is a very fine one. Good usually predominates in the long term, but often it is a close run thing. The moral of the story is: think for yourself, which can be done much more efficiently if there are no taboo or unmentionable subjects.

Finally, let us look at non-magical, non-religious irrationalities which relate to unusual phenomena rather than everyday events. The prime example in this category is the prevalent belief in flying saucers—the 'chariots of the gods'. This is not to be confused with the belief in unidentified flying objects (U.F.O.s) which undoubtedly exist and can be ascribed to all manner of things from weather balloons to meteorites and aircraft. But to explain these phenomena in terms of visitations by extra-terrestrials is irrational for the following reasons.

Intelligent life-forms are often asynchronous, i.e. they are separated in time as well as in space. Therefore any potential visitor not only has major problems in selecting the right point in space as a destination but far more serious difficulties in selecting the right point in time. Unless armed with prior information the alien would just have to take a chance that his arrival coincided with the existence of a contemporary civilisation. The life-span of a typical star is in the region of 10,000 million years, during which intelligent life is likely to exist for not more than ten million (it is highly improbable that man will still be here in ten million years' time). The chances of finding such life then, at any given moment, even on the one star in a thousand which is suitable, are more than a hundred to one against. The odds are that in the unlikely event of our star being selected for a visit, the aliens would arrive millions of years too early or too late and we would know nothing whatever about it.

Owing to the relatively few stars of any kind within 25 light years of Earth and the scarcity of suitable abodes for life out to a far greater distance, the overwhelming odds are that the nearest contemporary intelligent beings are at least 100 light years away—an estimate which would be regarded as wildly optimistic by most astronomers who have studied the matter. This means that journeys involving thousands of years of travel will be necessary, owing to the fact that relativistic mass-energy requirements† and interstellar gas and dust forbid travel at speeds even approaching that of light. Remember that even the most transparent regions of space contain a proton every few metres, and at high sub-luminal velocities these protons would take the form of a storm of deadly radiation.

The highest practical velocity is likely to be of the order of one tenth of the speed of light (which would still be about 20,000 miles per *second*!) and would have to be achieved four times over to allow for acceleration and deceleration at both ends of the journey. This would almost certainly

†see Appendix: Relativity

necessitate the use of anti-matter as a power source, and chemical rockets would be about as relevant as elastic bands. It follows that such technology is available only to civilisations very much more advanced than our own. Incidentally, you can forget any science fiction ideas you may have picked up about silicon-based life or disembodied electrical intelligences originating on totally unearthlike planets, as these ideas conflict with the laws of chemistry and evolution.

My somewhat optimistic estimate of an average 100 light years between galactic civilisations may seem a long way, but is in fact remarkably close, as it represents only one thousandth of a galactic diameter. In a typical photograph of a galaxy, two stars separated by this distance would appear to be almost touching, and it may well be that there are many such civilisations within the galaxy at any one time. Nevertheless, their average separation poses enormous problems for even radio communication, let alone actual physical contact, especially as there is no way our existence could have been detected at such distances. Our first powerful radio transmissions have reached out only about 50 light years into the galaxy, where they have become almost unimaginably feeble and difficult to detect. A few years ago the giant radio telescope at Arecibo sent an intentional message to a likely group of stars 24,000 light years away. If there will be anyone listening at the other end in 24,000 years' time we can expect a reply in almost 50,000 years (if there is anyone here to receive it, which is just as doubtful).

The mechanism of evolution by natural selection requires billions of generations to come and go before perfecting advanced intelligence, and this suggests that the lifespan of such creatures will not be so very different from our own, though lifetimes of several hundred years cannot be ruled out. It is not feasible that such beings should roam the galaxy for millions of years for the chance of finding intelligent life on perhaps every ten thousandth star visited. (At 20,000 miles per second relativistic time dilation is negligible, about half of one percent.)

To believe that aliens, having overcome all these obstacles and more, should then indulge in some ridiculous game of hide-and-seek with us must surely be carrying delusion to extremes. You may still think that such quaint notions are more a matter of ignorance than delusion or self deception, but if the natives of our island are ignorant of the facts what on earth have they been doing during eleven years of primary and secondary education? It would seem that aliens would be hard pressed to find intelligent life here!

The Loch Ness monster myth is a similar example of facts ignored and another ridiculous game of hide-and-seek. One has only to watch a negro chimney sweep searching a coal-cellar at midnight for a black cat which isn't there to realise that such games are characteristic of any search for an illusory being or creature. Certainly objects have been seen, probably tree-trunks and otters in the main, but to claim as some do that these objects are a plesiosaur-like reptile surviving from the age of dinosaurs is to flout common-sense.

Loch Ness was frozen solid during the ice age which ended about ten thousand years ago. How an air-breathing reptile could have survived this, or even the occasional surface freezing which has occurred since, is not explained. Neither (except by nonsensical arguments such as cannibalism) has the complete lack of tissue from dead specimens. The normal course of events with similar animals, even when they sink to the bottom, is that after a while the corpse is inflated with the gases of decomposition and therefore rises to the surface.

Why the creature should have chosen one of the smallest and least favourable bodies of water on Earth is also unexplained. One way to rule out the presence of large reptiles would be to observe the entire surface of the loch for a whole afternoon, perhaps from tethered balloons. Reptiles could be expected to break surface every hour or two in order to breathe. There are many other objections which I will not explore, but suffice to say there is no herpetologist or palaeontologist of any standing who takes the idea seriously, much as many of them would like to believe it. As with flying saucers, angels, ghosts and fairies there are many 'eye witness' accounts and dubious photographic 'evidence', but it is significant that alleged observers always forget their cameras when close up in conditions of good seeing. The activities of hoaxers cannot be ruled out, and there is no doubt that such people have played a significant part in perpetuating the flying saucer myth.

Archaeological mysteries of the ancient world such as the pyramids of Egypt or man-made geometrical landscape features in Peru are often explained in terms of visitations by extra-terrestrials, but the plain truth is that the ancients were cleverer than we sometimes give them credit for. As proof of that you need look no further than Stonehenge, which was definitely not built by astronauts and was already two and a half thousand years old when the Romans landed. Alternatively, you could say that in a little over five hundred years' time it will be two million days old, and if you look at a reconstruction you will see that it was a fairly sophisticated piece of engineering. Believed by archaeologists to have been a cross between a temple and an astronomical observatory, its function was probably to obtain information for astrological and religious purposes.

The idea of visitors from outer space arranging these marvels would be far more plausible if someone unearthed an ancient ray-gun or the equivalent of a 5,000 year old microchip. Alien astronauts are unlikely to be much different from the Americans in that respect, and probably leave litter and equipment behind them wherever they go. It is improbable that they examined the Earth without also landing on the moon, where their footprints would be preserved for millions of years. It is a virtual certainty that such footprints and artifacts will never be found, for the simple reason that they are not there.

The abominable snowman of the Himalayas and the sasquach, its North American equivalent, have a large army of believers who would have you believe that the creatures are relict populations of some ancient hominid (Gigantopithecus and Neanderthal man have been suggested). Gigantopithecus was exclusively tropical and sub-tropical, as were most other hominids, and would have avoided snow like the plague. He died out about half a million years ago. Some races of Homo erectus (Pekin man) and his Neanderthal descendants were cold tolerant, but their feet and therefore their footprints were identical with those of modern man, or at least of modern men who do not wear shoes. Pekin man died out about quarter of a million years ago and Neanderthalers about 35,000 years ago. Neither reached North America.

Film purporting to be of the sasquach shows a creature which looks nothing like Pekin or Neanderthal man, but suspiciously like a modern man dressed up in a monkey suit. The Americans always were great ones for hoaxes, and also great ones for believing them. Tracks of the Himalayan animals, always distorted by partial melting, might conceivably be of some new species of an already known animal, but are much more likely to belong to a known species of bear or langur which has ventured above the snowline. Some prints are probably of human origin but distorted by the hot sub-tropical sun. There is no more reason to ascribe them to an extinct hominid than would be the case for millions of other human footprints to be found in India or Africa which are indistinguishable (except, of course, by age) from those of fossil men. This is not a guess, as three and a half million-year-old fossil footprints have been found in Africa which look just the same as those of small modern men who have never worn shoes.

This brief survey of irrational beliefs is nowhere near comprehensive but merely a selection of examples. You could certainly add to them yourself if you gave the matter enough thought. The examples quoted are just those which sprang to mind and not necessarily the best. I have deliberately left out such lunatic fringe groups as Flat Earthers, believers in subterranean civilisations and so forth as individually their supporters are so few that they hardly count, though probably in total their numbers would be significant. It is hard to avoid the conclusion that our society has a craving for fantasy and illusion in preference to reality.

Exaggerated? I haven't told the half of it, and if anyone wants further evidence he should have no difficulty in finding it for himself. There may well be other islands of savagery in the world which are just as bad as our own or worse, but we do not have to live there. At least these other savages are not masquerading as Englishmen, that highly endangered species of renowned fair play and stiff upper lip, calm, civilised, reasonable, an honourable if slightly eccentric race whose word is their bond. Some will say that this archetypal Britisher never existed in the first place and is just one more figment of our fevered imaginations. Be that as it may, we can either make him exist or save him from complete extinction if we really want to. Not by wishful thinking or self deception, but by shunning the paths of savagery and promoting more rational and civilised attitudes in society.

"In the long run it is far more dangerous to adhere to illusion than to face what the actual fact is. And secondly, what is the point of life if you live in an invented world, if there is no relationship either to the world or people or anything. It isn't a relationship if you are related to something which isn't there or which is just there to make yourself feel comfortable. I think we all have this tendency and it is the problem of mankind . . . this is the thing which has always made mankind's life miserable and which is threatening our survival. We cannot face the implications of what we are doing and it will lead us over the edge into the abyss."

—Prof. David Bohm, world renowned physicist.

V

Information Processing for Biological Computers

THE IDEAS OF the vast non-realist majority only seem reasonable because countless hordes have been conditioned to think the same way. Not only does this demonstrate their lack of originality, but if their numbers could be reckoned in tens rather than millions even the illusion of reasonableness would disappear. If imitation is the sincerest form of flattery they must feel very flattered at their hordes of eager, unthinking admirers, though this following is certainly not justified by the merits or wholesomeness of their philosophy or the logic of their arguments. Seldom can flattery have been so thoroughly undeserved and imitation so unwarranted.

We now come to the crux of the matter when we consider the reasons for the kind of fuddled thinking we have been looking at. These are important because they show us the pitfalls to be avoided if we wish to see the world as it really is. Most of them can apply to otherwise intelligent people, and help explain why many such people hold irrational beliefs. The 'Adolf Hitler syndrome' would be a good name for the phenomenon, as the Fuhrer was an extreme example of how brilliance can sometimes go hand in hand with stupidity. These causes of irrationality represent impediments to sound reasoning, and as such are barely distinguishable from mental handicaps. Let us examine them in detail.

Low Intelligence. This is probably the only one which does not apply to people of average intelligence or above, and the least amenable to treatment. Where the deficiency is due to environmental factors, however, and concerns children, substantial improvements can be made if the problem is tackled early enough.

Ignorance, or Inadequate Education. This does not only apply to New Guinea tribesman but also to a great many natives of our island, who may have umpteen certificates to prove that they are 'educated' when in fact they are nothing of the kind. How can anyone be educated in this day and age when he does not even know what a star or an atom is, has no knowledge of evolution and the chemistry which makes it possible, and a mind packed with primitive superstitions and beliefs? However, unlike low intelligence, which is difficult or impossible to treat, this condition can be remedied. Childlike ignorance and naivety are only charming and attractive when found in the very young, where they are invariably accompanied by the childlike curiosity and open-mindedness which vanishes in later years.

Wishful Thinking. We now come to what is perhaps the greatest pitfall of them all. A great many things in life seem to be unfair. Not only seem to be but are. How much nicer if reality were more in accordance with what we would wish it to be. We really ought to be rewarded for virtue (but virtue often has to be its own reward, at least in the short term). It seems unfair that life is so short and death so final. It is almost unthinkable that any close relative or friend convicted of crime is not innocent, and when we commit crime ourselves it is usually amply justified. The true villain is someone else who has persuaded us, provoked us, or put temptation in our way. Those wicked people who want us to accept this rubbish they call 'reality' are certainly unpleasant and immoral, and probably mad into the bargain! In this way the trap is baited to lure the unwary away from truth. If we want to hit the target (truth), then like a skilled marksman aiming off for wind we should learn to allow for wishful thinking. As King Canute discovered and was able to demonstrate to his courtiers, wishful thinking is very ineffective as a means of changing reality.

Conditioning and Indoctrination. The human brain is a computer, and just like any other computer it can be programmed. Not only can be but must be. Even programmed to over-ride or bypass the normal procedures of logical deduction and reasoning. Unlike the majority of man-made computers, the average human brain is programmed to reject 'harsh', 'unpleasant', or 'disagreeable' facts and information. You can imagine what chaos would ensue if man-made computers were allowed the same latitude! If we are to perceive reality, all information must be assimilated, however 'harsh' or 'unpleasant' it might seem, subject only to checks for accuracy and reliability. The important criterion is not whether a particular piece of information is 'agreeable' or 'disagreeable' but rather whether it is true or untrue.

Chauvinism. A classical example is the proverbial 'male chauvinist pig', who maintains that anything done by a man must be superior to anything a woman can do. Scientific tests show that while men may be superior in some respects such as mathematical ability, women are (on average) superior in others—including, strangely enough, weapon handling, according to tests carried out by the U.S. Army. They are also said to be superior in verbal ability, though at risk of being a male chauvinist pig I suspect this may apply to only a small range of subject matter. Those who think that female chauvinist pigs are any better than the male variety are deluding themselves. We would do well to remember that something is not automatically better because it is a product of ourselves or our own social group. Black is not *automatically* beautiful any more than white is. Incidentally, it is a curious fact that virtually all books on birds of prey are written by men, and virtually all experts on this subject are white Caucasian males—something you can easily check for yourself by a visit to the library (minimum definition of an expert: someone with a lot more knowledge of birds of prey than just the falconry aspects and natural history of a dozen or so species.) So differences do exist, but whether these differences are a result of nature or nurture is an interesting question. There is a great deal of overlap in the abilities and aptitudes of different races or sexes. The scientific statistics refer to men and women in general and are not necessarily applicable to particular individuals.

The Wish to Belong to a Particular Social Group. This is probably synonymous with the 'Emperor's new clothes' syndrome. The acceptance of an individual by a social group is usually to a greater or lesser extent conditional on his acceptance of their outlook, ideology and beliefs, however crackpot and irrational these may be. The individual with an urge to belong thus has a powerful incentive to 'admire the Emperor's new clothes'. He will either persuade himself or be persuaded that the ideas of the group are reasonable and rational. Any evidence to the contrary will usually be rejected. An examination of lunatic fringe political or religious groups will tend to bear out this hypothesis.

Auto-Suggestion. The power of hypnosis is well known, and stage hypnotists frequently demonstrate their ability to convince their subjects of almost anything. Less well known is the phenomenon of auto-suggestion or self-hypnosis. This provides the means for an individual to persuade himself of almost anything, and is a probable explanation of the amazing powers of self-deception possessed by many people.

Mass Hysteria. This phenomenon is well seen in some of the more way-out religious ceremonies and political rallies in Africa, Latin America, and the Orient, but is by no means unknown in Britain. The attitude of mobs of young women to the arrival of a pop star, which is totally different from the way they would respond if they were to meet him alone in more sober circumstances, is one example. An interesting case occurred recently in the north Midlands, where nearly three hundred young people at a carnival collapsed or required treatment for nervous symptoms which at first led the authorities to suspect wind-borne contamination by pesticides or industrial chemicals. After a thorough medical investigation, however, the opinion of the experts was that the affair was due to nothing more than mass hysteria sparked off by a few cases of 'falling guardsman' syndrome. A number of similar occurrences have been reported in recent years.

Mass hysteria seems to have something in common with the wish to belong to or identify with a social group. The wish to 'identify' is a sign of weakness. To *understand* is far more important than to 'identify with' or 'relate to'. This weakness may well be responsible for the modern craze for sociology at the expense of more useful but intellectually demanding sciences. It would seem that many people have an inability to relate to anything more exotic than fellow members of the herd, and an almost total lack of interest in topics far removed from street corner and kitchen sink. Worse still, so acute are their powers of observation that most have completely overlooked the evidence which stares them in the face every day, and to which I draw attention in these chapters. Those who never venture outside the wood will not see it for the trees. Familiarity breeds not only contempt but blindness as well.

Fear. In this context fear takes several forms. Fear of rejection by social groups if we do not 'admire the Emperor's new clothes'. Or fear of the unknown, fear of what we might discover if we rejected fantasy and illusion and went actively in search of truth. It really amounts to cowardice and lack of moral fibre. The best remedy is mental self-discipline, which is not only the classic remedy for fear but is effective against most other obstacles to sound reasoning as well.

Lack of Curiosity, or the Perversion of Curiosity. Einstein once said something to the effect that he who is no longer curious about the universe around him is already half dead. Most children are naturally curious and forever want to know the whys and wherefores of everything—what the rainbow is, how far away the stars are, or where do babies come from. About the age of puberty this wholesome curiosity usually dies away—either naturally or perhaps suppressed by the dead souls and moronic intellects surrounding the average child. Its place is often taken by a perverted form of curiosity which is curious not about the wonders and mysteries which surround us but about the petty and mundane trivia of everyday life—"Who is so-and-so going out with . . . what colour is his car? What is my neighbour doing?" and so on. A very few individuals carry the childlike form of curiosity into adulthood, and are generally regarded as very odd fellows. Chimpanzees and other apes lose much of their playfulness and curiosity on reaching maturity, so this dulling of the spirit may be normal in the primates. This hardly makes it more attractive, and what is normal in chimpanzees is not necessarily desirable in man.

'Chatting Up'. Very few people stop and think of what is meant by the phrase 'chatting up', but when you do it takes on some alarming implications. Normally, it means telling your listener what you think he or she wants to hear rather than what is actually on your mind. Reality is avoided as far as possible in order not to introduce a jarring note by straining his or her grey matter, or drawing attention to some aspect of reality which the listener may prefer to avoid.

Disc-jockeys are accomplished exponents of the art of 'chatting up', which consists of talking as congenially as possible without actually saying anything. The aim is to soften up the listener in order to obtain something that you want. The term is usually associated with a persuasion technique commonly used on women, but in fact often used on men as well, especially those with rank and position and therefore favours to bestow. It follows that people with nothing anybody wants are seldom if ever 'chatted up'. When thus analysed in detail the custom is considerably less attractive and more sinister than it seems at first sight. It concerns us here because those people who are regularly 'chatted up' will have reality concealed from them to a greater or lesser extent, to be replaced by a comfortable cocoon of disc-jockey bonhomie and petty unreality.

Desmond Morris, in his original and thought provoking book 'The Naked Ape', describes a closely related form of verbal communication which he convincingly equates with the social grooming of monkeys and apes. In modern hominids it is no longer acceptable to search your companion's hair for fleas and ticks, so we have evolved a verbal substitute. With masterly understatement Morris describes this grooming talk as 'aesthetically unattractive', and cites our desire to pat and stroke other animals as a further manifestation of this grooming instinct. Whether or not Desmond Morris is correct, we could scarcely expect a more meaningful and profound relationship with animals owing to their lack of intellect, but with ourselves it is (or should be) another matter. The similarities between human behaviour and that of baboons (which, though less closely related than the apes, resemble us in being gregarious, ground dwelling primates not specialised for brachiation) are quite disconcerting, as any competent zoologist or anthropologist will tell you. This resemblance may be fine for the baboons but is largely outmoded in man.

'Brain Jamming', or saturation of the senses by obsessive preoccupation with a restricted range of subject matter (e.g. music or religion), especially 'pop' music. It is interesting to note that the Ayatollah Khomeini came to much the same conclusion quite independently, banning pop music on the grounds that it tended to blind people to reality but forgetting to mention that obsessive preoccupation with religion does the same thing.

There is nothing wrong with pop music in moderation, but carried to excess it can replace religion (Marx's 'opium of the masses') as a reality substitute, and is analogous to the jamming of a radio station. The brain as a receiver becomes so overwhelmed with spurious signals that its ability to take in more meaningful information becomes impaired.

It was perhaps not unreasonable for the wretched, down-trodden Chinese peasantry of the ninteenth century to seek refuge in opium, though the disgraceful role of British capitalism in supplying it (which finally led to the Opium Wars) cannot be so sympathetically treated. The boot today is on the other foot, and illegal exports of opium from the 'Golden Triangle' (Burma, Laos and Thailand), masterminded by Chinese warlords, are now helping to rot the fabric of *our* society. Opium is a substitute for reality, so anything which achieves the same effect is a substitute for opium in the sense implied by Marx, however harmless it may seem at first sight. If your existence really is as wretched and hopeless as that of a nineteenth century Chinese peasant—a highly unlikely contingency—you may have a valid excuse for seeking refuge in opium or one of its seemingly innocuous substitutes which will help you blot out reality.

Talking of opium substitutes, it has recently been discovered that there are naturally produced chemicals in the brain called endorphins, which have an effect similar to opium derivatives such as morphine and heroin. In fact, the reason opium and its derivatives have their effect in the first place is that they act on the very receptors in the brain which are intended for these naturally produced substances. Scientists have so far concerned themselves mainly with the pain-killing effects of endorphins, but it seems not unlikely that they may resemble morphine in other respects as well. It may turn out that at least some cases of hallucination are due to an over-supply of these natural neuro-peptides, and the stunted ability to hallucinate which is characteristic of total realists (including myself) may be the result of a deficiency! Many hallucinogenic substances are known and several are produced naturally in the body. Among the hallucinogens which produced religious experiences in the Middle Ages was ergot, a fungal disease of rye.

If we analyse the main objections raised by the many natives opposed to total reality, the superficial outlook and shallow perception of these people becomes apparent.

'Life would be dull because there would be fewer mysteries.'—It is true that as one progresses there are fewer mysteries, but there are more than enough to go around. A thousand human lifetimes would not be enough to solve all the mysteries that exist. The excitement of solving them and fitting the jig-saw together piece by piece is more than enough to make up for the dwindling number of mysteries. In any case, it is only when one investigates reality that one becomes aware of just how many mysteries there are.

'Life would be dull because if everyone accepted reality their opinions would all be the same, and there would be less to discuss and argue about.'—It is quite true that if everyone pursued the concept of rationalism there would be far less conflict and disagreement, but this is not necessarily a bad thing as it would mean less crime and fewer wars. There would always be sufficient differences of interpretation to provide ample material for discussion and rational argument, though in general there would be more harmony and accord. If one side argues that two and two make three while the other asserts that two and two make five, or even if only one side will concede that they make four, you have the seed of conflict and discord. On the other hand, if both will cast aside their biases and posturing in an honest attempt to perceive reality, then it must follow as the night the day that they will finish in closer agreement. "This above all, to thine own self be true. Then it must follow as the night the day thou canst not then be false to any man." What Shakespeare meant by this is that we cannot be honest with other people unless we are first honest with ourselves.

'People who prefer total reality and stark fact do so because they are lacking in imagination.'—Not at all. Quite the reverse in fact. It takes far more imagination to perceive reality than to accept fantasy and illusion in its place. According to the author Arthur C. Clarke, "Imagination is one of the first requirements of a good scientist". Einstein performed a brilliant feat of imagination and vision when he conceived the staggering idea that matter and energy are but two different forms of the same thing ($E = mc^2$)! His famous 'thought experiments' provide yet another example. It may truly be said that reality staggers the imagination.

Far from having greater imagination, it seems much more likely that non-realists have stunted and undisciplined imaginations which are incapable of visualising the world about them. Most have never even wondered what a star is and are quite incapable of original thinking. It was once said of an American president that you could wade through his deepest thoughts without getting your ankles wet. He was by no means alone. The uncontrolled and erratic mental fumbling of the average non-realist and his fuddled attempts to evade reality are no more to be confused with great imagination than the unco-ordinated thrashings of a lunatic are to be confused with great athletics.

'The stark facts of reality are mundane and unexciting.'—Another false assumption. There is an old saying that 'truth is stranger than fiction', and it really is. It is illusion and fantasy which are mundane and ordinary by comparison. To explain mysteries away as being due to magic becomes boring and repetitious after a while, as well as being dishonest. It is no accident that the people most hostile to the philosophy of total reality tend to be of low intelligence and so lacking in imagination that the only reality which has any meaning for them is that of their own immediate surroundings, rather like a dog, cow, kangaroo, or rabbit, or some other beast of the field.

'It is wicked to pursue reality if this involves conflict with the teachings of the Bible or is offensive to society.'—The truly wicked thing would be not to use the brain with which nature has endowed us or to turn our backs on destiny. A study of man's history shows clearly that it is human destiny to battle against ignorance and apathy: for better or for worse we are already too far along the road to turn back now. The die is cast. Three hundred years ago we burned innocent old ladies at the stake (so much for the Bible), whereas today we would give them meals-on-wheels and a home help. So we are making progress towards a fuller awareness of reality, but painfully slowly. Although exceptions may be found which prove the rule,† truth is fundamentally good and ignorance is evil. As for offence to society, those elements of society likely to be offended by reality are the bigoted, the ignorant, those with closed minds, and those with something to hide.

'Reality is cold.'—Not at all. It only *seems* cold at first sight, rather like a swimming pool when you dip your toe in the water. If you just have the nerve to take the plunge you will find that after the initial shock of immersion has worn off if is not nearly as cold as you thought and can actually be enjoyable. Come on in—the water's fine! This idea that reality is cold is a very common one. Siberia—that vast treasure house of wildlife and natural resources—also seems forbiddingly cold to people and animals who are not used to it, especially those brought up in some artificial and unnaturally sheltered environment. Should we therefore pretend that Siberia does not exist?

'Total realists are unpleasant people, devoid of warmth, feeling and emotion.'—Nonsense! They differ from other people mainly in not allowing emotion to cloud their judgement. Emotion is the enemy of reason, therefore the total realist must keep them separate as far as possible. But he is quite able to have both—there is no contradiction involved here. In a similar way sensible people do not put their luggage under the bonnet where it may foul the engine, but stow it away in the boot. In any case, the idea that emotion is exclusive to mankind and embodies all that is endearing and attractive in human nature just does

†Prove in the old fashioned sense of 'test'.

not stand up to examination. Not only is emotion found in many quite lowly animals, but even a sufficiently advanced computer could be programmed to feel emotion if we were prepared to accept the loss of efficiency involved. Make a list of emotions and note how many have innocent and attractive connotations. Precious few.

'Why go out of our way to find reality when it may make us unhappy?'— It is a gross oversimplification to say, as I have heard people do, that total acceptance of reality will make us unhappy, and the way to happiness lies in fantasy and illusion (and therefore by implication ignorance). This is the philosophy of the heroin addict. Although there is always a chance that some facet of reality will make us unhappy, the general effect of a wider acceptance of reality would be to reduce disagreement and unreason while increasing harmony and understanding.

However fervently you may wish to believe in a flat Earth, it does not alter the fact that in the final analysis you will have to live on a round one. Why not accept the fact? The Persian poet Omar Khayyam may well have been thinking along these lines when he wrote "The moving finger writes and, having writ, moves on: nor all your piety nor wit shall lure it back to cancel half a line, or all your tears wash out a word of it." So why not look reality square in the face and spit in its eye? Truth has a way of forcing itself on us in the long run, so if you have conditioned yourself to believe in illusion then reality will come as a greater shock when there is nowhere left for you to hide from it. The philosophy of the heroin addict is self-fulfilling. Reality is what it is, and all your fears, tears and wishful thinking will not alter it.

Reality is usually a paper tiger which may wear a fierce expression and look forbidding at a distance, but when seen close up you will wonder what on earth you were afraid of. It is when these paper tigers are not confronted and seen for what they are that the perversely self-fulfilling tendency of make-believe and delusion has a chance to turn them into something more sinister. The tragedy of Cambodia, for example, was not the result of some devastating natural disaster or ruthless invading army, but self-inflicted in pursuit of some hare-brained scheme to flout reality.

The end of the civil war in 1975, far from heralding a new era of reconstruction and prosperity, was the signal for the start of a campaign of lunacy and genocide which has few counterparts in modern times. The plan was to adopt a precisely opposite course to the one which has made another Asian country the envy of half the world, but the possibility that a precisely opposite course might achieve a precisely opposite result either did not occur to them or was considered a minor detail. Japan decided long ago to march boldly into the twentieth century, and has done so with a success that could scarcely have been envisaged even fifty years ago. Cambodia, having chased out the wicked Americans, decided to return to what they designated 'Year Zero'—an apt title in view of subsequent events.

Whereas Japan decided to adopt the technology of the West and has often brilliantly beaten us at our own game—yes, you've guessed it—

Cambodia, led by the Khmer Rouge and the maniac Pol Pot, decided it would be better to reject Western technology and return to something approaching the Stone Age in which virtually all machinery and technology (with the curious exception of military weapons) was abandoned! The cities were abandoned, even money was abandoned. Pnom Penh became a ghost town. The former urban population was driven into the countryside to do work for which their training was ill-suited and thousands died. Those who protested were imprisoned, tortured and murdered on a scale not known since the Nazi holocaust of World War II.

The people who told us this murderous government were all that was sweetness and light in contrast to the villainous Americans, and who now advocate a government of similar complexion for the United Kingdom, should be hanging their heads in shame. Should be, but needless to say, they are not. There is no shame in the never-never-land of make-believe, as it is specifically tailored to our fantasies. So you see, it is not reality itself but your attempts or the attempts of others to evade it which are likely to make you unhappy, and the dreams you wish to dream may turn into nightmares, not only for yourself (which is probably no worse than you deserve) but for other people too.

'Reality is too awful to contemplate. We need a little fantasy and make-believe to make life bearable.'—Well, here we are right in the middle of contemplating it, and up to now it hasn't been too bad, has it? Far from being too awful to contemplate, the exploration of reality is a great adventure which is open to anyone and calls for many of the qualities of more traditional explorers. There are untold wonders waiting to astound you if you will only make a little effort to look for them, and they are all the more enjoyable for being real. One has only to be shipwrecked on a desert island to discover that a real packet of hard-tack biscuits can be more delicious and certainly more sustaining than the most exotic imaginary banquet.

The trouble with fantasy and make-believe is that, like atomic radiation, there is no safe level. Even as you tell your little boy a fairy story, you are setting him up for con men in later life who would have him believe that there need be no relation between cause and effect and reality can be anything he wants it to be. I am not suggesting you should never tell your children fairy stories, any more than I would say you should never wear a luminous watch or sit in front of a television screen. As with radiation, we have to make a decision about tolerable levels where the benefits outweigh the risks. In both cases this decision is often somewhat arbitrary and, at least with regard to the make-believe issue, safe levels are less than commonly supposed. For proof of that you need only look at the catalogue of Cambodia-type disasters (as well as countless examples of a more everyday nature) which would have conservationists yelling blue murder if they resulted from radiation.

Certainly we need a little light relief, and even in the most enlightened society we could scarcely be expected to ponder the Theory of Relativity and similar profundities all the time. Those who can draw a sharp line between fantasy and reality can allow themselves a high ration of make-believe without much harm being done, but are paradoxically the people least likely to want it. We all think we can draw this line, just as we all think we can hold our drink, but are not always the best judges.

'If we insist on making them look at reality we might hurt peoples' feelings.'—The idea that the sort of needless suffering described in this book is preferable to the risk of hurting peoples' feelings is not only preposterous but warped and perverted. Hurt their feelings? Yes, we might, and not before time either!

'We all perceive the world in terms of models which are mental creations and not direct impressions of an external reality. Furthermore, these models are constructed from discontinuous electrical impulses registering in a discontinuous, cellular brain. What you see is not reality at all, but a mental fabrication which is no more real than anything I see.'—It is self-evident that external reality can only exist in the human mind in terms of models (it would be most uncomfortable to have a real bird or butterfly in there, for example, let alone anything larger!). The fact that these models are made up of discontinuous electrical impulses registered in a discontinuous brain is of little consequence. The astonishing discontinuity of three tiny spots of light painting fifty pictures a second in a colour television set produces a surprisingly faithful representation of reality. In any case, even external reality is discontinuous, being made up of atoms, molecules, quanta and sub-atomic particles. Another layer of discontinuity will not make very much difference.

More disturbing is the readiness of even the finest brain to fill in gaps in the available information with inventions of its own (hence optical illusions and certain other phenomena). However, not only do these inventions usually fulfil a useful purpose, but sufficient cerebration will distinguish them from genuine information resulting from the stimulus of external reality. Whether your view of reality is more reliable than mine depends partly on our respective abilities for this type of cerebration, though other factors such as honesty, emotional self-control and resistance to suggestion also enter the equation.

For the first time in history, technology and science provide us with the necessary pieces to put together a fairly complete and accurate picture of reality. The way these pieces fit together so perfectly, revealing an underlying order and symmetry in Nature is proof of its accuracy. Any false information will not easily fit in with the rest of the picture. The snag is that it requires a very broad range of interests to complete the jigsaw. An expert on nuclear physics, for example, will only have one piece (albeit in very sharp detail) and will require a very much wider range of interests if he is to put together a picture.

At the beginning of the previous chapter I mentioned some dramatic gaps in the knowledge of the average Briton. Gaps not in mere general knowledge but in knowledge fundamental to a proper understanding of the world around him. Anyone who thinks this can be cured by telling the individual concerned what a star is, or an atom, or about evolution or the

DNA molecule, has not really understood me. Not only are these gaps merely examples of a very much wider deficiency, but they are just surface symptoms of an underlying malaise which I also mentioned earlier—the dead souls and moronic intellects from which all vestiges of our healthy childhood curiosity have been lost. It is no use curing a few surface symptons without attacking the underlying disease, as any competent physician will tell you. The really lamentable thing is not so much the lack of knowledge as the lack of curiosity or any spirit of enquiry; not so much that they don't know as the fact that very often they don't want to know.

Somehow we need to revitalise their sense of awe and wonder, and promote an awareness of their surroundings which includes but also goes far beyond their own immediate environment. In most cases this appears to be a hopeless task which is not made any easier by the fact that many of their programmers, including some school-teachers, appear to be without these qualities themselves and can hardly be expected to inspire them in their charges.

"Why bother as long as we're happy?" the usual argument goes. Although superficially attractive the argument does not stand up to analysis. There is no good reason to believe that civilised man, for all his sophistication, is any happier than many savages, aborigines, or even monkeys. Rather the reverse in fact, if the suicide rate of countries like Sweden or the United States is any guide. All the evidence indicates that if happiness is to be the sole object of man's endeavours we might just as well still be swinging through the trees. So there must be some other reason for our existence.

Even if there were not, to build happiness on a basis of fantasy and illusion is to create a fool's paradise which depends for its continued existence on the evasion of reality and the willingness of other people to go along with the idea. The collapse of this house of cards is often the signal for frantic consumption of tranquilisers in order to shut out the world of reality which we have never acquired the ability to live with.

There is such a thing as destiny, and it is not a romantic invention of poets. It is a bird's destiny to fly because it has wings, and by the same token it is man's destiny to explore the boundaries of reality because evolution has given him the means to do so—his brain. For a bird not to use its wings or man not to use his brain would be a denial of their inheritance, though whether either in the long run will be any happier than their less privileged ancestors is debatable.

It is very unlikely that the exploration of Mars or Jupiter will contribute much to the sum of human happiness, but the reason man must reach out into the universe is the same as the reason for climbing a mountain or accepting reality—because it is there! For better or for worse it is our destiny to continue along the road we have trodden for almost three million years since we diverged from the apes. Any early men who turned away from that road have not survived, just as many birds which lost the use of their wings have become extinct.

Anyone who thinks I have exaggerated the prevalence of moronic attitudes in society need only look around him at the hordes of vandals and football hooligans which represent only the more noticeable tip of the iceberg. The fault does not lie with civilisation, but rather with the lack of it. It lies with the inadequate and retarded Neanderthal mentality which lingers on into the twentieth century where is has no rightful place. All too often man has the outward trappings of civilisation without possessing its substance, rather like those photographs of nineteenth century African chieftains or South Seas cannibals with top hats, rolled umbrellas, bare feet and grass skirts.

Further illumination will be provided if you question your friends as to their view of reality. The results will by and large confirm what I have said, although some ingenious evasions and red-herrings will be produced in the process. Having established the facts for yourself (don't just take my word for it) you may wish to question the list of possible reasons which I have suggested. All I can say is that if these are not valid and fail to hit the nail on the head I would very much like to hear what better ones you can suggest.

By now you may well be asking "What must I do to be saved?" The answer is fairly simple in principle but unfortunately a lot harder in practice. It means looking at things in a new light, as if you were seeing them for the first time (which in a sense you will be). It means finding and revitalising the almost forgotten remnants of that childhood sense of wonder and wholesome curiosity you thought you had lost for ever (and perhaps, alas, you have).

You will have to start by being honest with yourself, though perhaps being honest with other people will have to wait till later—let us not try to run before we can walk. The accumulation of relevant information which will enable you to define reality is an important part of the process but is not in itself sufficient (and I must here distinguish between mere general knowledge, which is desirable but not essential, and fundamental knowledge concerning the main issues which make up the pieces of the jigsaw). But the key to the whole issue is mental self discipline. Without it you are lost.

The majority of people use argument not to establish the truth but to establish a point of view—their point of view. The truth is often neither here nor there, and many are quite capable of inventing their own 'facts' as they go along. The important thing (as they see it) is to win.

The total realist, on the other hand, should be more interested in seeing reality than in winning some hollow 'victory' which, if he has sacrificed principles to obtain it, will actually be a defeat. This new way of looking at things will require considerable self-discipline from most people, particularly when they are not used to it. When fact, logic and reason indicate he is wrong, the total realist will accept reality. The pill is sweetened by the knowledge that he sees the world a bit more clearly than before. Unlike the majority of the human race, he must not cherish his illusions. Because they obscure his view of reality he knows he is well rid of them when they fail to stand up to reasoned argument. He must be prepared to butcher sacred cows (especially his own) with a dedicated ruthlessness wherever he finds them, in most cases quite without pity. This too requires mental self-discipline, especially as it is likely to make him rather unpopular. But don't worry too much about popularity—it is almost meaningless. Christ was so unpopular that he was crucified by the mob. Hitler, on the other hand, was extremely popular (inside Germany, at least) until about 1942, and more tears were shed at the state funerals of his thugs and henchmen than have been shed for many a saint. Many similar examples could be cited.

Although there is to some extent an order and symmetry in nature brought about by the laws of physics, the arbitrary and random nature of these laws is not designed to take into account human sensibilities. The result is that to our way of thinking many aspects of reality appear harsh, unfair or even unreasonable. The very unpalatability of this face of reality means we must exercise an unflinching self-control amounting almost to masochism when facing it, like soldiers going 'over the top' to walk doggedly into a storm of enemy fire. This is the real test of a total realist, which is why it is so important. If you can force yourself to face what you thought was unfaceable, to think what was formerly unthinkable, and accept the unacceptable, the rest will come easily.

Much of reality, as might be expected from the element of chance which comes into it, is innocuous in the sense that it seems neither harsh nor particularly gratifying to the sensibilities of our human nature. It therefore takes little strength of character or mental self-discipline to accept this aspect of reality, and still less to accept those aspects which we happen to find comforting, gratifying, flattering or otherwise enjoyable. To put it another way, if reality were always what we would wish it to be, composed only of the agreeable elements without a large component of the harsh or innocuous, we should immediately suspect that something was wrong, or at the very least that it was being manipulated by some thinking entity rather than the arbitrary dictates of the laws of chance, physics and nature.

The aspiring rationalist, however honest and sincere, will sometimes face a genuine dilemma in sorting fact from fiction. This is a problem of the utmost importance and significance, as any computer is only as reliable as the information fed into it. Feed nonsense in and you will get nonsense out. When programming our 'computers', therefore, we should be careful not to assimilate inaccurate information. This is very difficult to achieve, as we are constantly bombarded with a veritable Niagara of data, some of it accurate but much of it not. How to distinguish one from the other—that is the problem. Inevitably some faulty information will slip through, but with the application of a little common-sense and a few rules of logic and psychology this can be kept to a minimum and hopefully confined to subjects of little significance or importance.

Use reliable sources. This may be easier said than done, but it is usually possible to identify an authoritative source which is more reliable than others. Thus, if I want to know about crowned hawk eagles I would consult the well known Kenyan ornithologist Leslie Brown, but if I want to know about gyrfalcons I would consult the Americans Tom Cade or

Hal Webster. A source which has consistently shown itself to be unreliable in the past should be treated with caution, and preferably not used again. Even your own opinions, if they have been consistently wrong in the past, are no more likely to be correct in future unless some significant new factor is introduced into your reasoning process. Failing this, future opinions should be re-examined and compared with those of someone who has a consistent record of being right.

Cross check one source against another. If all sources agree, the information is usually reliable. If they disagree, then one source will be more accurate than another and further investigation will be needed. A source which conflicts with the majority can be suspected of being (but not assumed to be) unreliable.

Double check any source which is obviously emotional, biased, or has an axe to grind. Emotion, that great deceiver, is the enemy of reason, and it is human nature to distort, exaggerate, or even invent 'facts' to support their arguments. Conversely, those who readily concede some point which damages or runs counter to their own argument are particularly to be trusted, as such concessions are likely to be motivated by honesty and reasonableness. Their views should therefore carry more weight than the dogmatic pronouncements of those who are motivated solely by a desire to win.

Analyse and subject to tests of logic. Thus the peregrine is often quoted as attaining 275 mph (Mach 0·39!) in a dive, but analysis of aerodynamics shows that this speed would only be possible after a fall of 2,600 feet in a vacuum (unless we are to defy the laws of physics and introduce magic), plus a further 300 feet or so in order to pull safely out of the dive. There is no good evidence to show that the bird ever dives for much more than a thousand feet, let alone 3,000 feet in a vacuum! A ten-thousand-foot stoop from the top of Mount Everest might come close to the figure quoted but there is no evidence that the bird ever does this either. The extraordinary thing is that experts have been quoting 275 mph for years, without ever bothering to do the arithmetic!

Where two or more alternative explanations are available, choose the simplest of the alternatives which adequately explains the evidence. Or to put it another way, choose the hypothesis which makes the fewest assumptions and the minimum demand on blind faith. This principle (Occam's razor) is the cornerstone of the scientific method. Thus to say without any really convincing evidence that life originated elsewhere in the universe and was brought to Earth on a meteorite merely adds an unnecessary layer of complexity without attacking the underlying problem—the origin of life. It is simply an evasion, a red-herring which serves no useful purpose. As Newton wrote, ". . . Nature is pleased with simplicity and affects not the pomp of superfluous causes."

When, bearing in mind the limitations imposed by Occam's razor, you have several choices of information or explanations of roughly equal merit, choose the harsher or more disconcerting of the alternatives. This is to make allowance for our natural inclination to choose the soft option, the most agreeable. You are, so to speak, 'aiming off' to allow for wishful thinking. When the Viking space probe soft-landed on Mars in 1976 to conduct tests for the presence of microscopic life forms, the evidence was ambiguous and could be interpreted either way. In accordance with previous experience, the scientists selected the harsher, less welcome interpretation as their verdict. Although they may yet be proved wrong, it is extremely unlikely that they will be.

Beware of the obvious. Whilst there is of course a limit to the amount of checking and analysis we can do, there is a tendency to take external appearances and first impressions at face value when a slightly less superficial analysis would reveal a very different picture just below the surface.

Be suspicious of the inherently unlikely. This might seem at first sight to be so blatantly obvious as to go without saying, but there is something perverse in human nature which inclines us to reject the unvarnished truth in favour of far-fetched flights of fancy. It was the proud boast of Dr. Josef Göebbels (Nazi Minister of Propaganda) that the bigger the lie the more people will believe it, and he was probably right. He certainly spoke from experience.

Wherever possible gather information first and form opinions afterwards. To do things the other way around as many people do is to put the cart before the horse, and presents a grave temptation to tamper with the evidence so that it fits in with our preconceived notions. However formed, opinions are not sacrosanct and should not be regarded as immutable or unalterable. Be willing to change them when they no longer agree with the facts. We would do well to recall the words of Darwin—"I have steadily endeavoured to keep my mind free so as to give up any hypothesis, however much beloved . . . as soon as the facts are shown to be opposed to it."

Reject geocentric (Earth-centred), anthropocentric (man-centred) or egocentric (self-centred) views of reality. Each time man has produced a cosmology in which the Earth or man himself have occupied important and central positions, science has smashed it to pieces, and with good reason. One does not need to be an expert anthropologist to appreciate that such views are typical of savages, and contradict even the laws of chance, to say nothing of the laws of science. Primitive societies without exception hold anthropomorphic views of both animals and gods, but the very creatures which they see as semi-human or even gods in animal form are often little more than sophisticated automatons responding to stimuli. Despite their blood relationship with us, most animals do not see the world in anything like human terms.

As for the idea that we are in God's image and he in ours, you can imagine what an insulting suggestion this would be to any supreme creator unless he had a hearty sense of humour. The idea that we are the supreme life form in the universe and therefore in God's image is laughable. The very fact that we believe it is evidence of our lowliness. As intelligent life forms go, we are the fairly ordinary inhabitants of a not so very exceptional planet, and our morphology, far from being god-like, was dictated by such mundane considerations as the need to swing from tree to tree. There must be thousands of planets similar to Earth in our galaxy alone, and many life forms superior to ourselves will long ago have dropped such egocentric ideas, despite their better claim to god-like status.

There obviously is a supreme life form in the galaxy, but the chances that it is us are vanishingly small—probably thousands to one against. The sun is a very ordinary star (a yellow dwarf), and far from being at the centre of the universe is not even at the centre of the galaxy. Our galaxy, despite its marvels, is very ordinary and not particularly large as galaxies go. It is only one of billions, and occupies no important or central position in the universe, so you can judge the greater odds for yourself. The importance of even the most advanced intelligent organisms to the universe is negligible, though far in excess of our own. Our importance is not to the universe, which is unaffected whether we live or die, but to ourselves. Our purpose, if any, is of similar significance.

Our morphology was dictated by the particular circumstances of our evolution and other more advanced organisms will only resemble us in a very general way, sometimes hardly at all. Which of us was made in God's image and which are the bug-eyed monsters depends entirely on your point of view—beauty is mainly in the eye of the beholder.

It is admittedly slightly unscientific to accept the existence of creatures which we have never seen and are never likely to, but the alternative that the Earth is unique and its chemistry found nowhere else in the universe is far more unscientific, and breaks several rules of total reality. Anyone who thinks the kind of chemistry described in chapters one and two is unique to Earth is living in cloud-cuckoo-land. The only way this reasoning could be invalidated is if the laws of nature were found to be different elsewhere in the universe, but the evidence indicates that they are not.

Try to see things from as many angles as possible, particularly from the angle of an opposing viewpoint. This is similar to the military principle of putting yourself in the enemies' shoes and trying to see things the way they see them. It is only by viewing things from more than one angle that we are able to see them in three dimensions. The view from unusual angles will often produce new insight and understanding.

Look for paradoxes, anomalies and contradictions, and ask awkward questions. This is not to say that we can learn nothing from studying the normal, trivial and run-of-the-mill, the everyday and the ordinary. We cannot pick out the abnormal until we know what is normal. But we will make far more progress if, like Sir Isaac Newton, we devote the bulk of our energies to examining things which are in some way out of the ordinary and then asking questions which should be as awkward as possible without being ridiculous or frivolous. It is not generally appreciated that in any attempt to probe reality, and particularly in science, asking the right questions can be just as important as giving the right answers.

The total realist may compare himself with a scientist, and his view of reality to a new alloy for armour plate which he has invented. Instead of wrapping his brainchild in cotton wool or putting it in a glass case as many of his rivals will do, he should take it without delay to the nearest gunnery range where his ideas will be subjected to the most punishing tests. A few small cracks will not matter unduly, but if they fail the test completely—then back to the drawing board. Armour plate which needs to be wrapped in cotton wool is no good to man nor beast. He can at least afford to smile at his rivals' product, which they will take good care to shield from the smallest pea-shooter.

When researching the facts, try to avoid 'spotty' observations, which are usually the result of laziness. For example, you may find that after allowing for rose-tinted spectacles (as allow for them you must) your few close friends still do not resemble the rather unflattering description of the average human being given in this book. It may therefore seem to you that I am wrong and there is really little to worry about. We are not savages after all.

More likely your friends are atypical, and too few to represent a cross section of society as a whole. Check out a more substantial number, taking steps to ensure they are as representative as possible, be brutally honest with yourself, and see if you still come up with the same picture.

Arrange facts into an orderly sequence or pattern and the correct pieces should fit together like the pieces of a jigsaw. The pattern should be dictated by the facts and should not be an artifact of your imagination. Once a pattern is established, the existence of gaps can be detected and the likely form of missing pieces suggested. For example, no branch of science is complete in itself, but is inter-connected and reinforced by various subtle routes with other branches which usually support each other very well. Where they do not an error may be suspected, but this may be only an apparent error or contradiction which will be resolved by further investigation. Reality, by definition, can have no real contradictions, though it may sometimes seem to. Beware of forcing the pieces to fit—if you have the right ones in the right order they should fit easily enough.

Where there is difficulty in ascertaining truth, try to straddle the problem with intelligent estimates—a high one and a low one—then look for the true answer somewhere in between. To take the peregrine as an example again, the ease with which it overtakes fast quarry should make it obvious to anyone that the speed of the stoop could not possibly be less than 60 mph. On the other hand, a stoop from just over a thousand feet (the maximum, in normal circumstances) could, according to Newton's laws, barely attain 175 mph even in a vacuum. Make cautious allowance for aerodynamic drag and there is no way the upper figure can exceed 150 mph. So you have bracketed the target with fall of shot at 60 and 150.

Truth tends to lie midway between two extremes, so on this basis an estimate of 105 mph would not be far out even if a precise answer lay nearer the upper or lower figure. When you hear that terminal velocity for a human free faller is around 120 mph your confidence is increased. Some experiments with miniature speedometers have finally settled the matter, and indicate the maximum velocity of a very high stoop to be close to 100 mph, though this will vary slightly from bird to bird and according to altitude.

The problem need not be strictly mathematical for the bracketing principle to be useful. Even in non-mathematical problems one can usually bracket the target with a 'worst case' scenario which explores one extreme and a 'best case' scenario which explores the other. Remember too that inaccurate or irrational ideas when carried to their logical conclusion will usually result in absurdity. If the conclusion is absurd then the arguments that led to it must contain a fallacy, or at the very least a deficiency or inadequacy.

For example, consider this scenario: everything the so-called anti-realists say is true and they have brilliant imaginations. Check how well this fits the evidence. Now consider the opposite scenario: everything the so-called rationalists say is true and they are the ones with brilliant imaginations. How well does this fit? Now check the evidence again, keeping a wary eye for fraud and deception, and see if you can find a better fit at some point in between (though in this particular instance I suggest the truth lies nearer one extreme than the other).

Remember that reality is not always a matter of either . . . or . . . One alternative does not always have to exclude the other. For example, the meteorite catastrophe theory of dinosaur extinction does not rule out the climatic change theory or the collapse of food chain theory. It is quite possible that all three were involved (though I am personally very sceptical about meteorite catastrophe).

Is evolution gradual and imperceptible change which takes millions of years to transform one species into another, or is it a series of quantum leaps and sudden branching (saltation) whereby such changes are effected in only tens of thousands of years rather than millions? There is every reason to believe it is both and no reason at all to believe that one excludes the other.

Be suspicious of excessive tidiness. Because of the element of randomness and the imperfections of human nature, reality, in contrast to fiction, tends to be untidy with lots of loose ends. According to his butler, Gladstone's last words were: "I think I could eat one of Bellamy's meat pies." Which may or may not be true, but is rather more likely than "My country! My country! How I leave my country!"

It is said that Field-Marshal Montgomery would adapt and improvise to meet unexpected developments on the battlefield, and would afterwards claim that everything had gone exactly according to the prearranged plan. War, especially, is very untidy. The Field-Marshal should not have been afraid to claim the credit for his flexibility and resourcefulness, which are good military virtues. Be suspicious of excessive tidiness and pretty patterns which, especially where they concern the affairs of men, are the hall-mark of fiction.

Though human behaviour does tend to form patterns, in real life the element of chance, plus the erratic and irrational urgings of emotion, ensures a high degree of disorder with plenty of loose ends. Murphy's Law prevails. So if you think you see a neat and aesthetically satisfying pattern it may turn out that you have made a discovery like Mendeleev's table of the elements or Newton's laws of gravity, or you may have stumbled on a series of amazing coincidences, but it is more likely to be an artifact of someone's imagination. Not only is reality not cosy, but, except where it concerns the laws of nature, it is usually untidy as well.

Urge people not to take your word on trust, but to check for themselves. To say that something is so because it is convenient and desirable that it be so is just not good enough, so be prepared to give sensible reasons for your pronouncements. This will provide a strong incentive for you to get your facts right and take a more analytical look at your own arguments. When we are on shaky ground, the last thing we want is for doubters to investigate our flimsy foundations. But doubt is healthy, and making an effort to find the truth is healthier still.

The knowledge that they are investigating will compel you to build your argument on a rock. Not only that, but you will be favourably contrasted with those who tell us that doubt is a sin, that they are authority, and in no way should we question their fanciful dogmas. Human nature being what it is, when we think no one will check on us we are inclined to be lax.

When analysing and comparing, compare like with like as far as possible. The more unlike the things which are compared, the less valid the comparison, while to compare objects or situations which are totally unlike is not valid at all, and will result in false reasoning.

Reject the idea that it is wicked to question established thinking. Question anything which, after due thought and consideration, appears to conflict with the facts or defy logic. Above all, question assumptions. Remember, there are many people who, like Dr. Göebbels, have a vested interest in persuading you to accept some irrational idea or inaccurate information.

But why bother with information or information processing when it can all be done by intuition? There is such a thing as intuition in science, but in a very different sense from the generally accepted meaning of the word. Scientific intuition is governed by the sort of rules outlined above and is usually the result of subconscious calculation in a trained mind crammed with relevant data.

The accumulation of information, particularly what might be called 'fundamental information' as distinct from mere general knowledge, is essential to form an accurate picture of reality. It is largely from this perception of reality that our sense of values is derived; therefore it follows that those elements in society having a warped picture of reality will have a correspondingly warped sense of values.

This expectation is borne out by an examination of crime figures though, of course, by no means all non-realists will be criminals, so this is just the tip of the iceberg. Indeed, some will turn out to be 'do-gooders' who for understandable reasons tend to sympathise with criminals, in whom they recognise something of themselves. The extraordinary thing is that society requires us to accept dishonesty as a virtue, but only up to a certain point which is roughly that defined by law. The trouble is that some of us may not choose to draw the line at precisely the same somewhat arbitrary point as we are expected to. This, of course, is true of politicians, whom we are often and not without justification inclined to accuse of dishonesty. But a politician who was always totally honest would almost certainly never get anywhere—in other words, we demand dishonesty and deceit from them, just as we demand it from each other! The wonder is not that politicians sometimes overstep the mark, but that they do not do so more often.

The opposite of reason is unreason, and one of the striking features of most crime is the unreasonable and mindless nature of the criminal mentality. A common feature of the criminal outlook is a preoccupation with material things and a tendency to evaluate everything in terms of money, to which they attach an exaggerated importance and are sometimes prepared to go to extraordinary lengths to acquire quite trivial sums. But remember, criminals are only carrying the more acceptable 'virtuous' dishonesty, which most of us cultivate, to its logical conclusion by refusing to recognise our artificial boundaries. The mindlessness does not start at those boundaries but long before we reach them.

Can there be the slightest doubt of a connection between standards of honesty and perceptions of reality? Can there be any doubt at all that the 'respectable' mindlessness, which most of us support, breeds the more vicious kind which has no respectable facade? And can there be any doubt that the mindless majority fear the cure more than the disease and will therefore allow the suffering to continue more or less unabated? Time and time again dangerous criminals are described in court as 'Walter Mitty' characters who have difficulty in distinguishing fantasy from reality. We tackle the problem at the wrong end, after the Walter Mittys have left the factory, whereas we should start by destroying the production line which is turning them out.

An individual's perception of reality is also dependent to some degree on intelligence, and as people have different interpretations of the word it will not be out of place here to attempt a definition. People possessing a good memory (and especially those who are able to pass exams on the strength of it) are often said to be intelligent. It is a fundamental misconception to confuse memory with intelligence, otherwise preposterous situations present themselves. For instance, we should have to regard a tape-recorder as intelligent. Whatever else intelligence might be, I think most people would agree that it is not an attribute of tape-recorders, however instantly they might remember the most obscure of scientific lectures and despite their ability to repeat those lectures word perfectly perhaps twenty or thirty years later. So we must ascribe intelligence to some other characteristic which is not possessed by tape-recorders.

There are thought to be two kinds of memory—short term and long term—which are possibly controlled by two different memory systems within the brain. The short term system is arguably even less credible as a manifestation of intelligence, though of undoubted usefulness in passing examinations. Genetic memory or instinct, of the sort which enables a bird to build its nest, is incorporated in the structure of the DNA molecule and therefore expressed as proteins and enzymes, but just how far this is analogous to the mechanism of memory as such remains a matter for conjecture. It may be that all systems are dependent on establishing unique patterns of electrical connections in the circuitry of the brain, and that the inherited proteins and enzymes are somehow able to bring this about.

Quickness of learning is also widely regarded as synonymous with intelligence, and whilst it is true that more advanced brains (man) will learn more quickly than a primitive brain (tortoise) we cannot allow the validity of this definition either, as learning is largely (and at times entirely) a function of memory. It is only to be expected that advanced brains will excel in this respect, as the quality of all their systems is generally superior, including information storage and recall. It is debatable as to whether quickness of any kind is truly a measure of intelligence. Imagine two computers, one built for speed and the other for accuracy. On being fed with similar data one comes up with an almost instantaneous answer of dubious accuracy, whilst the other, though taking ten times as long, generally gives answers of much greater accuracy and reliability. Which computer is the more intelligent? This, as I say, is perhaps debatable, but I think as regards utility the slower but more accurate answer is preferable except in situations where time is of the essence.

I think we are agreed then that whatever intelligence is, it will be of characteristic of brains and computers but not of tape-recorders. So what do brains do that tape-recorders do not? They reason, compare, analyse, deduce, compute. And here I think we have it. The ability to reason is intelligence, and as we might have expected is not found in a tape-recorder.

A useful insight into the working of the human brain will be obtained if we compare it to a man-made computer. Indeed, the brain is no more than an organic or biological computer working on an electro-chemical system as opposed to purely electronic. The main difference (up to the present time) is one of complexity. The largest computer yet built is comparable in complexity and ability to the brain of the smallest ant, but we can expect to see considerable advances on this within our own lifetimes.

The reason a computer may appear to be more intelligent than an ant is that it is specialised for a particular function. An ant's computer has to be all singing, all dancing, all purpose, controlling a thousand and one different functions from analysis of sensory organ information to breeding. This computer is programmed almost entirely by heredity. Indeed, as already mentioned, it is thought that memory, or information storage, may be encoded in the structure of a molecule, analogous to the information encoded in the DNA molecule, nature's own 'computer chip'. The smallest ant, by the way, is barely larger than the full stop at the end of this sentence. Its brain is microscopic.

By contrast, the human brain is programmed by both heredity and environment—mainly environment: nurture rather than nature. Anyone having influence over a child is in effect a computer programmer, but whereas a long course of training is deemed necessary in order to program the comparatively primitive man-made computer no such training is required to program a child, with results that you can see for yourself all around you. The average child is appallingly badly programmed, hence his stunted awareness of reality in later years and a prevalent tendency towards mindlessness and vandalism.

The relatively small variety of 'programs' available to children in this country is almost frightening, as they not only tend to produce an 'admass', stereotyped, production-line type of human being churned out by the million, but the inferior quality of some of these programs produces a mindless, moronic type of individual. Anyone who thinks I am talking nonsense had better look around them again, or just pick up a newspaper, this time with their eyes open.

Part of the blame lies with our schools. Some of these establishments are actually prepared to say out loud that the willingness and ability to conform to the mindless charade which passes for reality in our society (the acquisition of social skills, as they would put it) is far more important than mere academic attainment, which might taint impressionable young minds by acquainting them with reality.

Supposing we wanted to conceal reality: how might we go about it? Firstly, we could divide it up into small bits and show only one piece at a time (do this with a picture and if the pieces are small enough your audience will not fully understand what they are looking at, especially when the pieces are jumbled and not all of them are shown). Secondly, we could magnify the trivia and the superficial detail of our immediate surroundings until it filled the space left vacant by the missing reality, leaving no obvious gaps. Thirdly, we could condition our audience to believe that reality should be emotionally satisfying, and to find it repulsive when it is not. We might also suggest that only wicked, unpleasant people would want to go looking in out-of-the-way places for awkward facts and credibility gaps. This will ensure that they are satisfied with the compartmentalised view (the small pieces) and the magnified domestic detail which we have substituted. For good measure we could magnify the occasional small piece as well.

Something like this seems to have actually happened, though whether by accident or design is hard to say. Probably a bit of both. The information dealt with in this book, though of fundamental importance to understanding the world about them, is either not taught at all or taught in such a compartmentalised way that it is difficult for pupils to relate any one piece of the jigsaw to any other, and therefore to the whole. This recalls the deliberately compartmentalised way in which spy networks are organised: individuals are given just enough information to get by, but not enough for them to put together a clear picture of the whole organisation.

In the case of education it may be that this has happened just by chance, or even incompetence, rather than any more sinister motive, but it happens so often that there are grounds for thinking it is sometimes deliberate rather than accidental. It seems quite probable that there are educators who do not want their pupils to question the status quo and deliberately deny them the means to do so. Whatever the motivation, the fact remains that the matters dealt with here are a closed book to most school leavers, or for that matter most adults. Why? If pupils were educated to be able to take part in intelligent discussion and informed debate instead of making soothing noises and body language to each other like twittering starlings on a chimney pot, the world would be a lot better off. In fairness to some high class schools and competent teachers, it should be said that they realised this a long time ago: but not enough of them, as many parents are only too painfully aware.

The analogy with starlings is particularly apt, for starlings tend to be very unoriginal and slavishly imitate one another instead of thinking for themselves. When one flies they usually all fly, when one turns they all turn, one feeds and they all feed. One chooses a roost and a million others choose it too. It may seem almost poetic justice, and certainly helps prove my case, that some of the establishments criticised above are burned down by the very products of their own shoddy assembly line. Arson is overtaking conkers as a pastime for schoolchildren.

Before leaving the subject of intelligence, there is some evidence that both memory and the ability to reason improve with practice and training. Memory, though not in itself intelligence, is nevertheless an important sub-system—where would a computer be without its data bank? It follows that the brain, like the body, will become degenerate with lack of use. Just as the average person's body is degenerate, so that he is incapable of running more than a few hundred yards (and that with risk of a heart attack), let alone five or ten miles, so his ability to reason is degenerate and incapable of breaking free from the strait-jacket of his programming. If tackled early enough, however, there is some hope of improvement in both cases.

Again, just as good diet and adequate nourishment is essential for the proper development of the body, there is good evidence that it is also essential for the proper development of the brain, especially as regards protein intake. This, as well as their inferior programming, accounts for the apparently lower intelligence of certain races and social groups (to mention yet another taboo subject), though other reasons may be involved as well. Given a good diet from infancy and quality programming with adequate mental stimulation and training, these people would probably turn out to be just as intelligent as anyone else. The self-fulfilling character of theories which justify racial discrimination on grounds of inferiority is obvious.

Whilst the amassing of large quantities of data (a function of memory) is not a direct measure of intelligence, nevertheless there is a tendency for more advanced brains to gather information. They recognise that it is the raw material for their raison d'être, almost like a fuel or a food. A computer is nothing without information, so its accumulation will be at least a sign of intelligence though not an accurate measure. As is well known, curiosity, which is connected with information gathering, is more developed in the larger-brained animals.

The trouble with intelligence tests is that they are based on a set of cultural values and preconceived ideas which are not common to everyone. Experts admit that I.Q. rating is more an indication of a person's ability to do intelligence tests than an absolute measure of intelligence. In the interests of objectivity, however, it should also be said that most of those who score high marks on such admittedly biased tests would also do fairly well on any honest and reasonable alternative which might be devised.

If it is possible to produce a reasonable-looking test which discriminates against certain races or social groups, then by the same token it should be possible to produce tests which favour them without seeming to be irrational or obviously biased. On the other hand, no one with an exceptionally low I.Q. could possibly obtain the opposite result with another test unless that test was so biased as to seem ludicrous. To suggest that these races and social groups are not less intelligent than others, despite severe social and economic handicaps, is equivalent to saying that they would actually be superior once these handicaps were removed.

The freedom of choice which most people believe they have is largely an illusion. By and large they will think only what they have been programmed to think, and this in turn will dictate their actions. The average person has never had an original thought in the whole of his life! Indeed, he would be most offended if you suggested he was somehow different or unusual. This is one of the many paradoxes of human nature, as no one wants to believe he is merely a stereotype, churned out on some moronic mass-production line, but at the same time few wish to be thought of as unusual or different from their fellows in any but the most superficial characteristics. In this latter ambition most of them succeed admirably. Conformity for conformity's sake is generally accepted as a virtue. Even television educational programmes have extolled the virtues of being ". . . the same but different", while making it abundantly clear that the sort of differences they have in mind do not extend to much more than finger-prints. Pressure to conform is often so subtle that victims—which means just about everybody—are unaware of it.

There is a pathetic belief on the part of many individuals that an accurate perception of reality will somehow come to us by some sort of divine revelation (intuition is a word sometimes used), without the slightest necessity for any actual work or effort. This belief is naive in the extreme, especially as most of the people around us have a vested interest in concealing reality from us and would prefer us to believe in all sorts of irrationalities, from ghosts, spiritualism and flying saucers to miracles and astrology.

If only a fraction of these weird superstitions were true the world would be a very odd and nonsensical place indeed, a sort of nightmarish Mad Hatter's tea-party. The looking-glass world of Lewis Carroll and other writers is only enjoyable because we can abolish it at any time and return to reality just by closing the book. We are conditioned from an early age to believe that an entirely unedited version of reality is unacceptable or even unreasonable, and that others have a moral duty to enter into the spirit of make-believe which we or our group will substitute for it. Remember how Alice tired of the charade and dared proclaim: "You're nothing but a pack of cards!" When we too have shed the burden of pretence we feel a similar sense of freedom. The Emperor really does have nothing on . . . what a relief! We are not going mad after all.

The pursuit of reality demands hard work, not only in the accumulation and analysis of data but in the mental training and self-discipline involved. It really needs to become a way of life—you cannot effectively be a part-time realist, switching from rationalism to apathy and hallucination as it may suit you from day to day. If you are not prepared to make the effort but prefer to go along with the crowd, so be it, just as long as you do not delude yourself that you are seeing anything close to reality. You would not claim to be a marathon runner unless you were prepared to put a lot of effort into the training, would you? But supposing that you are tired of being a lemming, or a member of some mindless pack, herd, flock, mob, or similar social group, and would like to pinch yourself awake from the Mad Hatter's tea-party in order to see things as they really are, how do you go about it? I will now set out the requirements of the aspiring total realist:

He must first of all recognise that he does not yet see reality, be tired of living an illusion, curious as to what the world is really like, and willing to make the effort to find out. He must be willing to scrap his existing 'program' or at the very least drastically modify it. This reprogramming is essential.

The first step—and perhaps the easiest—is to actively acquire accurate data on the universe which surrounds him, and after checking for accuracy and reliability (cross-checking different sources is one way to do this) he can then allow his computer to assimilate the information. Information which is even slightly dubious should be first held in a state of limbo rather than swallowed immediately, hook, line, and sinker. Only after further checks have established its authenticity can it be confidently fed into your database. It will soon become apparent how neatly the facts fit together, like the pieces of a jigsaw, and this in itself becomes a useful test for accuracy. False information will not easily fit into the overall picture. It will also be seen that a credibility gap exists between the picture that emerges and the one most people would like us to believe in. (The word 'culture' is only a conventional way of referring to computer programs for advanced biological computers. Think about it!)

Most people seem to think that whatever seems to be obvious at first glance represents reality, especially when it is what they want to believe—a trap which the total realist must be careful to avoid. Whether fear, laziness or dishonesty lies at the bottom of it, there is certainly a character weakness involved somewhere. Perhaps such people are afraid to turn stones for fear of what they might find underneath, or perhaps they are just too lazy to dig any deeper. For example, it is 'obvious' that total realists must be lacking in imagination and are hard, emotionless, unfeeling individuals. It should be realised that truth seldom gives itself up without a struggle, like a partridge leaping ready plucked into the arms of the hunter, begging to be cooked. Rather it must be remorselessly hunted down, tracked through wild and arduous terrain and flushed from its hiding place, while many a self-appointed gamekeeper will try to ensure that you come home empty handed.

The aspiring rationalist must recognise that it is human nature to substitute wishful thinking for reality, and be on guard against it. When several alternative interpretations present themselves, seemingly supported by equal amounts of evidence, he should choose the harsher or more disconcerting of the choices left to him, while allowing for the scientific method of choosing the simplest hypothesis which will adequately explain the evidence. This is because he recognises that wishful thinking will spoil his aim, and he must therefore 'aim off' to allow for it, like a skilled marksman aiming off for wind. The very fact that you would like to believe something is good reason for regarding it with suspicion. Many people would like to believe that I am talking nonsense, for example! Do not believe your emotions—half the time they are lying to you.

He must try to see things in a new light, almost as though he were seeing them for the first time, as up to now he has probably seen only what he was programmed to see and therefore what he wanted to see. He must be aware that he and the people around him have been conditioned to see things in a certain way—indeed, the term 'brainwashed' would not be an exaggeration in some cases.† We often admire things not because they are admirable but because we have been conditioned to admire them, and what we are accustomed to see as virtues may not be virtues in their own right, but only appear to be because we have been conditioned to see them as such. This explains why some most peculiar ideas and dubious morality are commonly regarded as virtuous when in fact they are nothing of the kind. If we step back to look at them in true perspective, it is apparent that virtue, like beauty, is often in the eye of the beholder and will vanish when viewed from another angle, like the illusory colours caught in a morning dewdrop.

The would-be rationalist must not allow fear to deflect him from his course. Reality is not nearly as cold or unpleasant as most people imagine it to be—it is more a question of getting used to it. Neither should he be over concerned about popularity. As I have already said, it is almost meaningless—usually just a sop to our vanity and self esteem. Some of the nicest men have been very unpopular, and some of the most evil have been very popular indeed. Given a choice, people usually prefer to do or say whatever is popular rather than what is right, fair, or reasonable—another frailty of human nature.

†See Appendix:Suggestion.

Many people evaluate everything—even other people—in terms of money ("What use is reality if you can't make any money out of it?")! If the aspiring realist happens to be one of these, the practice must cease forthwith. Few things of fundamental importance, and fewer people, can be evaluated in terms of money. A certain sex whose identity I will not reveal—I prefer you to judge for yourself—are particularly guilty in this respect.

If the aspiring rationalist is a male chauvinist pig or any other kind of chauvinist, he should try and mend his ways. He must use argument as a means of arriving at the truth and not a battle to make a point of view prevail at all costs, regardless of fact and logic. Above all, he must never invent 'facts' to support his views or present as facts information which he suspects is unreliable. A 'victory' obtained by these means will actually be a defeat.

He must not be willing to follow the herd regardless of where it might go. It can happen (though admittedly not often) that the rest of the regiment is out of step, and not the individual. At the Mi Lai massacre in Vietnam, for instance, only about six people refused to take part. The rest wanted to show that they were 'one of the boys' and were not 'cowards'. To refresh your memory, the incident occurred in 1968 when a company of American troops (about a hundred) marched into a Vietnamese village and virtually razed it to the ground, killing about three hundred† civilians at point blank range, mostly women, old people, and children. When U.S. helicopters came in to take out the survivors, they had to train their guns on their own troops, many of whom were 'high' on drugs and behaving like animals. That much-sought-after likeness to the majority which most of us take so much pride in is rather less flattering once you have been deprived of your rose-tinted spectacles. The 'virtues' we have for so long admired turn out very often not to be virtues at all, but manifestations of weakness, stupidity, dishonesty and mindlessness.

Although the boy in the story was in fact spanked and sent to bed without any supper (which is not generally known) the total realist must likewise refuse to conform with the 'Emperor's new clothes' syndrome. He must also harden his heart to slaughter sacred cows wherever he finds them, especially his own. Occam's razor may be wielded with deadly effect. He should stay his hand only for those few individuals so utterly dependent on their delusions that they could not survive without them. It is most unlikely that they would listen anyway, but the total realist is not inhuman or heartless, and both war and justice should be tempered with mercy. One does not kick the crutches away from under a helpless cripple, but these should not be confused with the vast majority of malingerers who are only suffering from a slight deficiency of moral fibre.

He should beware of 'chatting up', especially if he is possessed of goodies which others might want. (One is here reminded of Aesop's fable about the fox and the crow.) He should ask himself, "What is this fellow after—what does he want out of me?" Shakespeare probably had this facet of human nature in mind when he wrote the story of 'King Lear'.

†The official Vietnamese figure, which may well be correct, is 601.

The two daughters who 'chatted up' their father were only out for what they could get, and having got it were no longer interested in him. Cordelia, the apparently ungrateful daughter who refused to butter him up, turned out to be the only true one in the end. Beware of people who only tell you what you want to hear.

Both Aesop and Shakespeare were great students of human nature, and were trying to tell you in a round-about way what I have put more bluntly. In those days people had to watch what they said, on pain of death. They had no option but to make subtle hints. As you can see, the centuries of heavily camouflaged suggestion and innuendo have not achieved very much, but they have, I hope, created a climate where a more direct approach becomes possible.

The student of reality must not swamp his information receiving apparatus with an overload of meaningless signals by becoming preoccupied with a restricted range of subject matter. Instead he should try to broaden his outlook so that he will acquire as many pieces of the jigsaw as possible.

Pop music addicts please note—the reason (if it had not already occurred to you) that disc-jockeys are totally absorbed in pop music and appear to live for little else is because it is their livelihood. They depend on it for their bread and butter and many of them make a small fortune out of it—you do not. Therefore it is unwise to accept their evaluation unless you intend to become a disc-jockey yourself.

Anyone who has qualms of conscience about where our conditioned dislike of reality has led, is leading, or will lead, should try to revitalise that almost forgotten childhood curiosity about the world around him and revise his ideas about what is interesting. He has probably been programmed to believe that the trivial, the commonplace, the frivolous, the mundane, the petty and superficial aspects of his everyday existence are by long odds the most interesting. Even what someone had for breakfast, though no different from what ten million other people had, will be found interesting.

Sir Isaac Newton, arguably the greatest scientist of all time, likened himself to a boy playing on the seashore ". . . diverting myself in now and then finding a smoother pebble or prettier shell than ordinary, whilst the great ocean of truth lay all undiscovered before me." Notice that it was the more attractive shells or unusual pebbles which were found interesting. The ". . . great ocean of truth" which poor Sir Isaac found himself unable to explore so well as he would have liked in the seventeenth century is today fairly well charted.

All too many modern Britons, heirs to the noble traditions of Newton, Faraday and Darwin, may be likened to someone panning in a stream, throwing away the gold dust but keeping the sand and gravel. You have a chance to explore reality which Isaac Newton would have given his right arm for. Don't throw it away.

"If I find something interesting," the argument goes, "then it must be interesting. There is no way to define whether one thing is more interesting than another, because all views are equally valid. What is interesting is merely a matter of opinion." Not quite. There are ways of sorting the wheat from the chaff and defining what is truly interesting from the rest. Put yourself in someone else's shoes: ask yourself "Would a Martian, or even an intelligent Russian, be interested in this if he was here at this moment?" The fact that there are no Martians or that Russians have their own brand of make-believe and opium substitutes is beside the point—the essence of the idea should still be apparent.

Another test is similar to the test of great art or literature—the test of time. "Would this be of interest to someone in ten, fifty, or a hundred years' time?" you might ask yourself. Also, "Is it the sort of thing which Australopithecus might have found interesting?' There is little doubt that Australopithecus would have very much agreed with the more superficial point of view. (Being rather dim witted and in any case preoccupied with survival he was an extremely bad judge of such matters, though a great expert in body language and social skills. Perhaps with a broader outlook and more enquiring mind he might have lasted a bit longer.)

Though admittedly rather rough and ready, this method will usually sort out that which is inherently and fundamentally interesting from that which is not. After experimental verification of his Theory of Relativity by Sir Arthur Eddington in 1919, Einstein became something of a celebrity and members of the public, including well known figures of the day, were clamouring to be introduced to him. To his dismay, however, he found that the vast majority were interested not in his revolutionary new theories but in such mundane domestic trivia as the colour of his socks. "Are they mad or am I?" asked an exasperated Einstein!

The total realist should have no taboo subjects which he is afraid to bring out into the open in case they cannot stand the light. Uggh! Nasty! Unhealthy! Leave that sort of thing to the 'New Victorians', the latter-day prudes whose sensitive souls require all sorts of apparently innocent topics to be regarded as unmentionable, especially the subject of reality. Bring it all out into the open where the sun and air can get at it, and if the light of day reveals imperfection and decay we will then be able to take remedial measures. Like the prim Victorians, there are many people about today who believe that too close an examination of reality is unhealthy and likely to reveal parts which should remain hidden. Just who are the real perverts, the lookers or the concealers, is something you can probably figure out for yourself. The total realist must not cherish his illusions. He must see them for what they are—unnecessary impediments and handicaps which obscure his view of reality. He is well rid of them.

He must not play with words as many of his detractors will do. This is one of the favourite types of red-herring used to distract attention from the issue debated. Non-realists tend to be devious characters who, when confronted with reality, will attempt to avoid it. They will go round it, under it, over it—anywhere but straight at it, and seem to see nothing dishonourable in attempting to sidetrack or dodge the issue.

I remember once using the word 'incredible' in its colloquial sense of 'extraordinary'. The other fellow insisted that the word meant 'unbelievable' and that his statements were not unbelievable. In this he was technically correct, but as he knew quite well what I meant his ploy was just a meaningless exercise in semantics, playing with words. The important thing is the intended and understood meaning, not some irrelevant red-herring to be used as a substitute for rational argument. Leave that sort of thing to the rabble. Any reasonably intelligent person will understand the ideas which I am trying to convey in this discourse, so let us have no petty minded hair-splitting about the meanings of words. It is the ideas which are important though, of course, we should make an effort to express ourselves clearly.

If you remark to a non-realist that the universe is self-sustaining and works well enough by itself without any need of God, he will very likely retort that it has no need of you either. The salient fact that God is claimed to be an indispensable creator and manipulator who runs the whole show whereas no such claims have been made for you seems not to worry him. It is not that he is too stupid to have noticed this point, but dishonesty and evasion have become a way of life with him, and he cannot resist the chance to keep himself in practice. Like Bishop Wilberforce of Oxford Debate and Darwin fame, his artificial arguments are notable for their superficiality and deviousness rather than honesty or any sound basis in fact or logic. Such people are an excellent example of what not to imitate.

Finally, a dedicated seeker after truth should put far less priority on being the same as everyone else and much more priority on being rational and reasonable. Towards this end he should try to think for himself instead of accepting or rejecting what he is told to suit the philosophy of the herd or the whim or dictates of his emotions. In the words of Descartes: "I think—therefore I am!" There is nothing wrong with conformity as long as it is not just conformity for conformity's sake. Conformity to truth, high standards or ideals is in no way forbidden to the total realist. Conformity to mindless mediocrity is quite another matter. The notorious blunders of bureaucrats would largely be avoided if they sometimes paid more attention to what is rational and reasonable rather than the mindless gobbledegook of their forms and rule books. Perhaps they are examples of memory masquerading as intelligence. It should hardly need saying that to take a contrary view just for the sake of being contrary, or because a contrary view is emotionally more acceptable, is not the same thing as thinking for oneself.

"Working hard to think clearly is the beginning of moral conduct."

—PASCAL

VI

THE VAMPIRE EFFECT

TO MANY PEOPLE the idea of accepting total reality is anathema, as we have seen. Their aversion can be quite alarming in its irrationality, revealing as it does a side of their personalities normally kept hidden. When confronted with the issue, the very idea seems to make them ill. Some will visibly shrink away in horror like claustrophobiacs threatened with confinement in dark cupboards, or recoil like vampires in the presence of a crucifix. Even as they do so, they will try to keep up the pretence that they have nothing against reality and have not been conditioned to find the subject unpleasant.

In a sense they may even be telling the truth—they have not *knowingly* been conditioned. This is the insidious thing about programming the human computer. Usually the subject is unaware that he is being conditioned, and even the programmers themselves are likely to be unaware of what they are doing. Which is perhaps why they often make such a mess of it.

Reactions to the subject of reality (and perhaps reactions to this book) would make an interesting and informative psychiatric test. In between total rejection and total acceptance would come a whole spectrum of human conditions. Such tests would be useful for certain jobs, such as the Army, where rigid adherence to reality and the ability to define it is important. It would be instructive to examine the reactions of psychiatrists themselves—would they favour some degree of rejection or would scientific reverence for truth compel total acceptance? On an insanity scale rated from 0 (completely rational) to 100 (certifiable) no one would score zero! Total realists would have very low scores, probably in single figures, with most people clustered around the mid way point and a fair number edging up towards a hundred. Criminals, lunatics, rationalists, anti-rationalists, psychiatrists, we are all insane, though not necessarily certifiable. It is simply a question of degree. This is why insanity evidence at criminal trials is often controversial and ambiguous. There is no sharp boundary where normal mental aberration leaves off and criminal insanity begins. A lot of 'normal' people are walking time bombs, just waiting for the right circumstances to trigger them off. It has been seriously suggested that the ability to hallucinate in times of stress is advantageous and was favoured by natural selection.

With even the best of intentions reality may sometimes be overlooked or go unrecognised in error. The total realist cannot claim to be always right, as this claim would itself be a denial of reality. What he can claim with every confidence, however, is that as a searcher for truth who has disciplined and dedicated himself to understanding things as they really are, he is, all else being equal, more likely to be right than someone who has not.

Much more reprehensible than honest error is wilful evasion of truth by refusal to hear the evidence. This has become a regrettable feature of our society in recent years, and even in university debates speakers are sometimes shouted down by student 'intellectuals'. This is a gross violation of the principles of total reality, as is the notion that certain aspects of reality should not be explored in case the results should not be to our liking. The self-censorship imposed of their own free will by most individuals to shut out some aspect of reality is bad enough, but censorship imposed from without to any substantial degree has been a hallmark of totalitarian regimes and corrupt societies throughout history. Given the ignoble and sinister tendencies inherent in all bureaucracies, some aspects of our own Official Secrets Act are enough to cause disquiet, and have prompted attempts to introduce a Freedom of Information bill. This has predictably been opposed by the forces of darkness which bureaucracy seems to cultivate, and which thrive like a mildew in totalitarian states. Among the less savoury things for which Britain is renowned is its secretive form of government, which in this respect is almost unique among Western democracies.

Censorship, which is essential for the distortion of truth and concealment of reality on any large scale, was widely practised in Nazi Germany. In addition to the loss of freedom of speech and freedom of the press, book burning was a prominent feature, and the books of authors such as Einstein, of whom the State disapproved, were consigned to the bonfires. In such a situation, honourable men felt almost insulted when their works somehow escaped the flames. To the hordes of critics who banded together to shout him down, Einstein replied that if he had been wrong then one dissenting voice would have been enough. His own attitude to censorship was clear and unequivocal, as might be expected from a man of science: "As long as they are not violent I want to let everyone say what they wish. For myself, I have always said exactly what pleased me." The Nazis were also responsible for promoting various forms of pseudo-science which served the needs of politics but had very little to do with reality. It speaks volumes for human nature that the ideas of the vast majority on censorship are more in line with those of Dr. Göebbels than they are with Einstein's.

The denial of reality is a denial of truth, and hence a form of dishonesty. It says much for the duplicity of which the human mind is capable that this does not seem to bother the vast army of non-realists, many of whom consider themselves to be honest men and women. At the root of the problem is emotion, a good servant but a bad master which all too often is allowed to rule our minds.

Emotion was originally introduced by nature to enhance the survival prospects of various organisms by replacing reason in situations where their limited intellect was not adequate. Love for a mate or offspring has inspired in many a bird or mammal the appropriate action to fend off danger. Fear lends wings not only to birds. Anger has enabled animals to rout foes they would normally give way to, in order perhaps to recover stolen food or otherwise aid survival. Emotion, a non-specific and highly versatile form of instinct, is only with us because it has survival value and has therefore been favoured by natural selection.

But animals do not reason in the sense that humans do. Because their brains are so much smaller and less well organised they have no intellect in the usual sense of the word; therefore the fact that emotion is the enemy of reason is no handicap at the level of reason available to animals—far from it. Very often it makes an excellent substitute. With humans it is another matter entirely. Because of our vastly superior intellects emotion no longer makes an acceptable substitute, and while it would not be true to say that it has outlived its usefulness it has nevertheless become a handicap in some respects. This state of affairs was never intended by nature, which originally designed emotion as an asset, to take over where intellect and the more specific forms of instinctive behaviour were inadequate. The deficiency in intellect has now largely been made up, thus rendering the system outmoded and obsolescent. Emotion is now in search of a role to replace the one it is losing.

One possibility for this new role may be likened to the exaggerated perspective and false colour used in the interpretation of military air-reconnaissance photographs, or the exaggerated relief sometimes used in maps or graphs, giving at once both a heightened while at the same time a distorted picture of reality. This seeming contradiction is not really a contradiction at all, which is explained by the fact that the interpreters of maps and photographs are aware of the distortion but are willing to tolerate it for the more vivid and useful perception of reality which it gives.

By this analogy emotion, which is within our power to control but not abolish, may be useful at times in order to heighten our perception of reality, provided we are aware of the distortion produced. It also retains part of its animal function in that it is still useful as a motivator, despite the fact that we are often thereby motivated to do things which would have been better left undone.

Einstein was once asked by a friend whether he believed everything could be expressed scientifically. "Yes," he replied, "it would be possible, but it would make no sense. It would be description without meaning—as if you described a Beethoven symphony as a variation in wave pressure." He did not mean by this that we should ignore reality, or that emotion and its handmaidens, dishonesty and deceit, made an acceptable substitute. What he meant was that the finer emotions, in reasonable amounts which are just sufficient to do the job, can sometimes give more meaning to our perception of the world around us. The baser aspects of human nature, against which Einstein fought all his life, have no place in this role. Einstein had many of the attributes of a total realist, and was always careful to deny any suggestion that he was a mystic. His wife too, on these occasions, would say "Albert? Mystical? You must be joking!"

The universal acceptance of reality would bring about the greatest social and political revolution the world has ever seen, to the great benefit of mankind. Imagine a world where the actions of both nations and individuals were dictated by reason (not to be confused with expediency) rather than emotion or self-deception. By extrapolating into the future already discernable trends it will be seen that something close to this desirable state of affairs is almost inevitable in the long run, provided we do not blow ourselves up in the meantime. Whether the time-scale involved is of the order of hundreds of years or of thousands depends on you and others like you, but let us have no illusions that this revolution can be effected on a wide scale within our own lifetimes.

Perhaps the quickest and most effective way to bring it about would be to introduce a new subject into the school curriculum—rationality training, i.e. training in the principles of total realism. (Something very similar was recently introduced in Venezuela, where a Minister of Intelligence is responsible for a scientific programme of intelligence stimulation in the very young, and pupils are required to attend regular 'thinking lessons' based on the ideas of Dr. Edward de Bono.) It is very unlikely that the authorities could be persuaded to initiate this training within the forseeable future for several reasons. For one thing, in order to make way for the new subject something else would have to go—perhaps pop music appreciation or some similar pseudo-academic form of time wasting of the sort currently practised in our schools.

Another reason is that the establishment relies to some extent on the lemming mentality in order to manipulate the masses, despite the inconvenience to itself when some non-establishment individual or organisation does the same thing. My own opinion is that the advantages of a totally rational society would outweigh the disadvantages for the establishment. Imagine the benefits if a Chancellor of the Exchequer were able to reason with the electorate in order to bring about some economic objective, instead of fighting a losing battle to prevent the lemmings committing industrial suicide. By the same token, however, he would have to be much more honest with them (and with himself) when seeking election or re-election, as a non-lemming, thinking, reasoning, truth loving electorate would be far less easily fooled.

Some people may raise the objection that, subject to certain qualifications, there are perhaps more important things than an accurate perception of reality—a point of view with which I feel inclined to agree. What these people overlook is that reality and these other values are by no means incompatible. We are not faced with a choice between one or the other. Far from being mutually exclusive, an acceptance of reality will tend to reinforce the finer aspects of human nature, and any inhibitory effects will strike hardest at the negative aspects of our personalities.

Remember that the 'warm, lovable human beings' who cannot do enough for you and whose lack of comprehension you are so willing to excuse are often erratic, motivated by unfettered emotion and mindlessness, and would just as readily have your guts for garters if the spirit moved them (as it sometimes will). The people around you may not be as benevolent as they seem, as you may have already discovered to your cost. The evidence is overwhelming. Inside many a civilised Dr. Jekyll

Above: Contemporary print of Conquistadore atrocities in South America.

Below: Bodies litter the ground after the mass suicide at Jonestown, Guyana, in November 1978. A similar event involving only about seventy people occurred in the Philippines in 1985.

Above: Adolf Hitler—monster or kindly father figure?

Below: Nazi book-burning, Berlin, 1936. Several other countries have also tried book-burning, but most have more subtle methods of suppressing the truth.

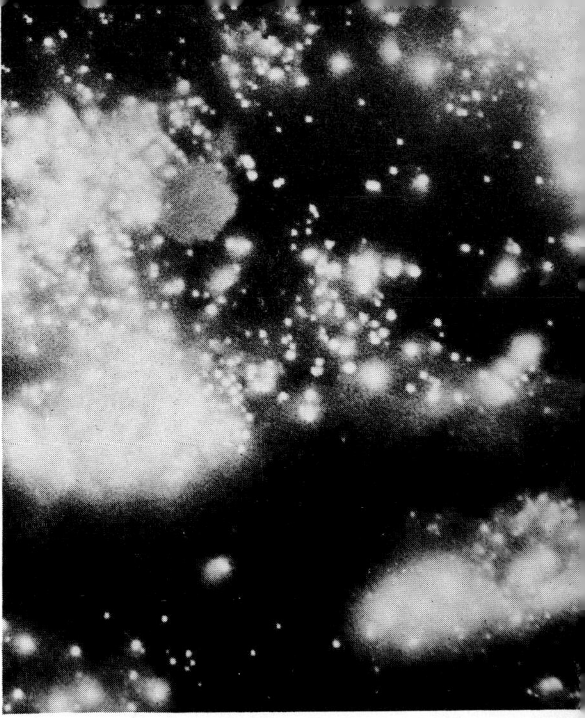

Above: Political prisoners massacred by the Khmer Rouge. Between two and three million people were murdered by the Pol Pot regime in Cambodia between 1975 and 1979.

Above right: The firestorm rages in Dresden's first and last air-raid of the war. The updraught created howling gales at street level, sweeping fleeing civilians into the flames. Eighteen square miles of the city were destroyed.

Below: Dresden after the air-raid of Feb 14th 1945. As many as 135,000 may have died. There were no military or industrial targets of any significance.

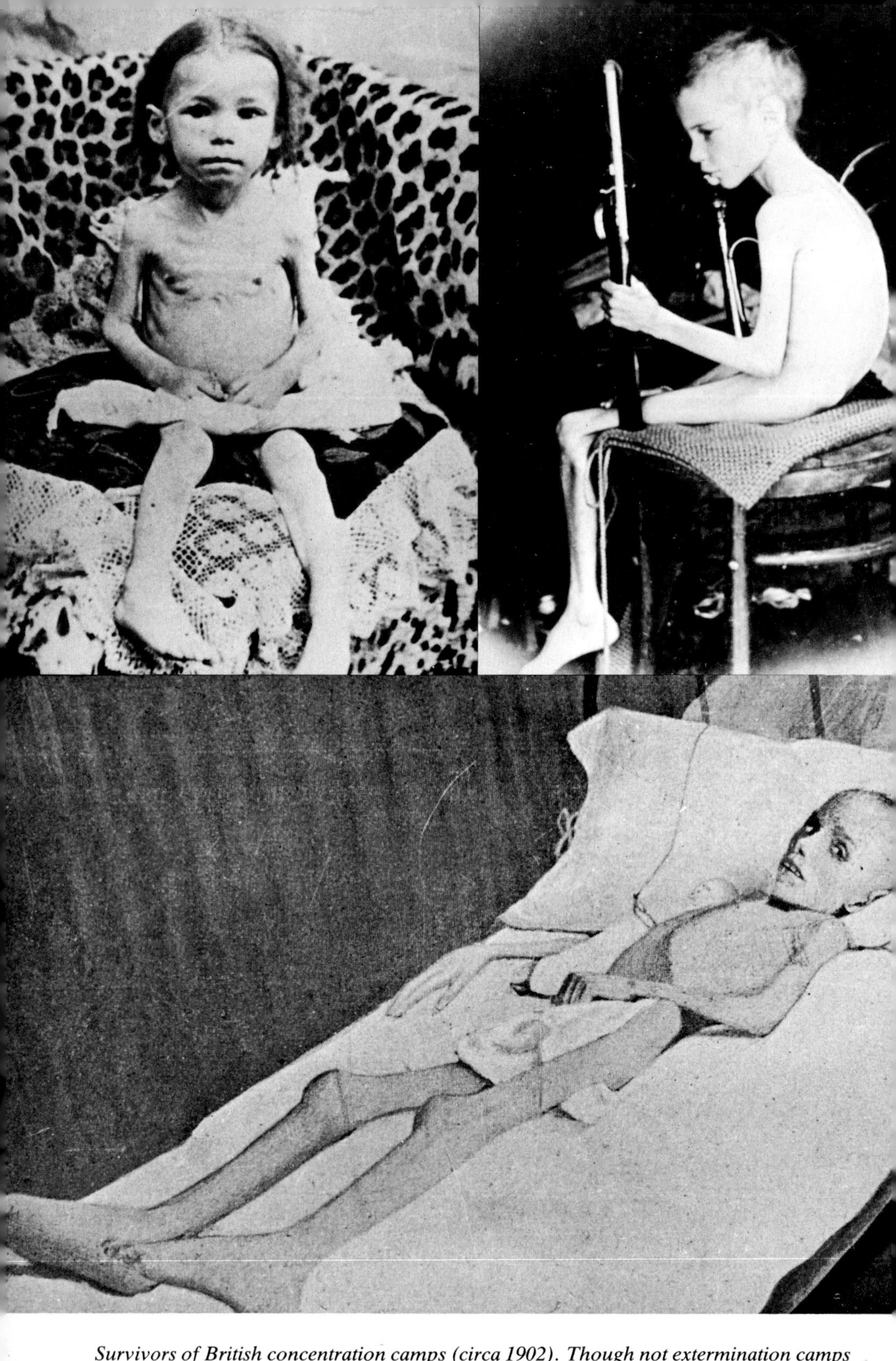

Survivors of British concentration camps (circa 1902). Though not extermination camps as such—neither was Belsen—and inmates died mainly from malnutrition and disease, this was little consolation to the victims. Nearly 30,000 died. "Terrible things happen in war," say the anti-rationalists mysteriously, as though that justified everything.

lurks an irrational Mr. Hyde, nourished by mindlessness and screaming to be let out. Feed him well enough and he will burst his bonds.

Most of the animals prowling the asphalt jungle are warm, lovable human beings for most of the time. Many are only restrained from committing the most barbaric crimes by the thought of the penalties involved should they be caught, and an ever rising percentage are not even restrained by that. Do you really want to be like them? The honest reasonableness, however 'cold', of the average rationalist will provide you with a far better protection than the heartwarming but mercurial kindliness of the non-rationalist and over-human majority.

There is an inherent morality in the philosophy of total reality which transcends the emotional and therefore perfidious morality of religion, as well as being more solidly based. Reality, by definition, is not founded on fiction and is unable to resort to the bribery and blackmail which are a feature of most religions. Fairness, for example, (or even-handedness as the Americans call it) although a man-made notion which is unknown in nature, is strongly suggested by the concept of symmetry which occurs time and again in the natural world, particularly in the laws of physics. See how relativity encourages us to look at the other fellow's point of view!

This in itself would recommend the idea to total realists, but much more important is the fact that without an even-handed approach truth becomes almost unattainable. And without liberty, freedom of speech and the printed word, the total realist is gravely hampered in his quest, so fairness, honesty, toleration, love of liberty and truth are just some of the virtues (or evils, as some would see them) which spring spontaneously from an acceptance of reality.

The idea that evasion of reality is just a harmless and amusing eccentricity or an endearing quirk of human nature is not borne out by the facts. In late 1978 a group of American religious fanatics hiding from the world in their jungle camp at Jonestown, Guyana, were visited by a United States Congressman who was investigating complaints about their methods. Ironically enough, most of these complaints were from anxious parents who had unwittingly set their children up for just such deception, brainwashing and exploitation by conditioning them to participate in the make-believe world around which society is built and which, eventually, most people are unable to do without.

Suspicions proved to be well founded and, threatened with an end to their unwholesome dream world, members of the cult first murdered the Congressman and four of his group of aides and newsmen (seriously injuring ten others), then committed mass suicide by taking poison. About a thousand died. It all started out with the 'harmless' ideas with which they were programmed as children. A journey of a thousand miles begins with a single step. Mindlessness promoted for whatever virtuous purpose will sooner or later be put to the service of wickedness. Such disasters (and worse) are an annual event, but for various reasons many go almost unreported. James 'The Amazing' Randi, the world famous magician who has campaigned against illusion masquerading as reality and was one of the first to recognise the dangers, remarked at a recent meeting that every demonstration of the so-called paranormal contains the seeds of the tragedy at Jonestown. He might also have added Auschwitz, Dachau, Belsen and other abominations of the Third Reich, the torture chambers and execution squads of Cambodia and Idi Amin's Uganda, and a thousand and one similar avoidable disasters of modern times.

Hitler, incidentally, though never a house painter named Schickelgruber, homosexual or syphilitic as many non-realists would have you believe, was a great believer in astrology and other half baked ideas. It would not be correct to brand him as stupid, however, and in other ways he was brilliant—some might say a genius. He was certainly not a reasonable man, but could be charming when it suited him.

So common are the manifestations that there are times when one might be excused for thinking there was some sinister conspiracy to suppress awareness of reality and promote the charade of make-believe which usually takes its place. "All the world's a stage . . ."—and how! There is, of course, no conspiracy as such, but what does exist is this: a general understanding (which for obvious reasons is seldom if ever put into words) not to rock the boat, not to prick the bubble which we have been conditioned to believe is the only alternative to hell on earth.

In reality the reverse is true—it is make-believe masquerading as reality which brings living hell for millions. For them the dream has become a nightmare, as narcotic dreams are all too apt to do. We despise the heroin addict who must have his 'fix' and to hell with the consequences, but are we any better ourselves? And in the long run can the outcome be any less appalling? Make-believe is just like any other addictive drug: because of people who abuse it the happiness it brings is far exceeded by the misery caused. The history of human disaster is largely a story of injury self-inflicted in the pursuit of illusion and self deception. The world has been taken over by a sinister force but no one has noticed, for the name of the sinister force is human nature. It carries the seeds of our own destruction.

The boat needs to be rocked, and its sleepy, self-indulgent occupants thrown into the water to wake them up. The unwelcome cold bath will actually be a kindness, for a boat crewed by the sort of people I describe will sooner or later drift over the precipice; it is only a matter of time—not a question of whether but a question of when.

I have been putting considerable emphasis on the word 'reasonable', and the reader will very likely be wondering "What is reasonable, and what has it to do with reality?" The outline of the answer should be apparent in what has already been written, but I will endeavour to explain more fully. As has already been remarked, the opposite of reason is unreason. Define the unreasonable and what is left will probably be reasonable. Since, as we have seen, unreasonableness is the hallmark of the non-realist and therefore contaminated, it should, in the absence of a suitable barge pole, be held at arm's length with a long pair of tongs. We must take care not to become infected.

To take an historical example, it was *unreasonable* for the Conquistadores to murder, rape, pillage and deceive in the name of Christ, as even to an unbeliever such as myself it is clear that Christ would have been appalled. The way his philosophy of gentleness and unselfishness was so easily stood on its head is typical of the way people are able to make black equal white whenever it suits them. Their treacherous murder of the Inca ruler Atahualpa, having asked and received a room full of gold as ransom (and having previously massacred his entourage), is a deed that will live in infamy, though far outdone in wickedness by some of their other crimes and atrocities. The rationalisation was that the Inca's bodies, gold, or their very lives were trivial repayment for the salvation of their immortal souls.

Just as the Nazis were able to celebrate the birth of a Jew while spending the rest of the year murdering as many as possible, the contradiction in what they were doing seems to have escaped them. As to whether a bunch of Bible-spouting cut-throats and criminals is qualified to redeem anyone's soul (should there be such a thing) or whether they might have been better occupied in saving their own is a matter I leave you to judge for yourself.

It is unreasonable to demand such qualities as loyalty, truthfulness and honesty from others if we are not prepared to honour these values ourselves. To put the idea in its most mechanical and unemotional terms, it is a question of maintaining symmetry. Parity is seldom violated in nature. Observe the symmetry in "Do unto others as you would that they should do unto you"—a concept, by the way, which did not originate with Christianity.

Mindlessness in all its forms is unreasonable. The human mind is our only claim to superiority over the animals. In most other respects they leave us standing. There are many animals which can outrun, outswim, outfight, and outjump us, animals whose agility and reflexes make us seem like blundering dinosaurs, and whose vision, hearing, sense of smell and touch make ours seem pathetic. We have only one thing which no other animal can match: the finest biological computer the world has ever seen. To shamefully abuse or neglect it is a denial of our inheritance and likely to lead to our extinction.

It is unreasonable to refuse to look at evidence because we suspect that it may prove us wrong (a self-indulgence which we are careful to deny our man-made computers), or to ignore an avalanche of evidence that we are wrong in favour of the most feeble suggestion that we may be right. At the Battle of Copenhagen Nelson put the telescope to his blind eye, but he was not ignoring the evidence. He was ignoring a line of flags which represented the opinion of someone more remote from the battle than himself, and subsequent events soon proved him right. So do not imagine when you ignore evidence that you are displaying the Nelson touch.

It is unreasonable to see what we want to see rather than what is actually there. It may be argued that because it is sometimes *expedient* to do these things, they are therefore not unreasonable. Some people might go further and say it is reasonable to do anything which is to our own advantage, as reason is to do with fact, logic and deduction rather than morality. Like many half truths this seems at first sight to be a powerful argument, but withers before the onslaught of analysis and the weight of the evidence. If you look through what I have written you will see that an accurate view of reality requires an almost masochistic intellectual honesty, the will to press on regardless of pain in the relentless pursuit of truth. There is no place for cissies and they soon drop out. As truth is elusive and many people have a vested interest in trying to hide it from you, its pursuit is not usually a matter of expediency—rather the reverse. Expediency, like emotion, is the enemy of truth.

In the Soviet Union, Rumania, and probably other countries, idealists voicing truths which are unpalatable to the State (and here the unacceptable face of bureaucracy rears its ugly head again—or is it the unacceptable face of human nature?) are deemed insane and locked away in mental institutions for 'psychiatric treatment', a modified version of what went on in Nazi Germany. There is a strange and almost convincing Mad Hatter's logic to this attitude. "Surely," it goes, "people who contradict the State knowing full well that their lives will be made very unpleasant indeed, if not snuffed out completely, must be mad." But it is characteristic of non-realists to confuse logic and reason with expediency. The dissidents of the Soviet Union are not lunatics but very brave men and women. Those who cannot see the difference perhaps deserve our pity more than our contempt.

You can imagine what happens to total realists in a totalitarian state, unless they are willing to keep their mouths shut. This latter imposition comes hard to people who believe that truth is its own justification. What I have written would have been more than my life was worth even in England four hundred years ago, and if the more hard-line anti-realists had had their way, the same would be the case today. Many Englishmen if fact were cruelly done to death for doing nothing more reprehensible than stating the truth.

The most courageous men are sometimes forced by the enemies of truth to abandon their principles in favour of expediency. The great Galileo, under threat of death by burning as an heretic, agreed to renounce his discovery of the moons of Jupiter and proof that the Earth goes round the sun. But even as he made his retraction, we are told, he muttered defiantly under his breath ". . . and yet it does!" His deference to expediency in this case is quite understandable, and I hope that in his place I would have had the good sense to do the same. It is unreasonable to expect others to accept martyrdom in defence of truth, although they may sometimes choose to do so. (Another illustrious Italian, Giordano Bruno, was burned at the stake some thirty years earlier for vaguely similar offences.)

To those who choose expediency as a matter of course I say this: think again. Expediency and self-interest are not always one and the same, although they may seem to be at the time. Apart from anything else, it is a matter of enlightened self-interest for us to promote rational attitudes and reasonable behaviour in the community. If, for example, we vandalise a telephone booth, it may feel good at the time, provided our personalities are sufficiently warped, but in the long run we all have to pay.

You may argue that many vandals are not apprehended and neither do they pay rates, therefore the entire cost is borne by someone else. Is it? Apart from the fact that the vandal debases himself as a human being his parents very likely pay higher rates, taxes, and other charges as a consequence of the rising tide of vandalism, and this in turn affects his standard of living. His grocer and other traders also pay higher rates and insurance as a result of vandalism, and this is passed on to the customer in higher prices. In business as in life, nothing is free. From trading stamps to social relationships, there is always a price to be paid, however indirectly.

Perhaps one day the vandal will need to make an urgent telephone call and find the booth vandalised. Perhaps his imitators will feel the need to release their inhibitions when passing his car or some other property or personal effects. After all, if it feels good—do it! So who is paying now? This applies not only to vandals but to all those who reduce the quality of life by their unreasonable behaviour. Many of these people are quite popular among their social companions, not because of any particular physical or mental excellence, but because of their willingness to conform even to mindlessness and mediocrity. Mindlessness has become elevated to the status of a virtue, and decadence and degeneracy are seen as desirable social assets as long as they do not differ too much from the decadence and degeneracy of the group.

There is something in human nature which makes most people want to shout from the house tops "Hey, world, damn you—this is me. I exist!" even when the claim is nothing to boast about and wiser men would prefer to keep it a secret. This urge is manifested in the harmless but at the same time mindless way in which lemmings—sorry, people—delight in having their names read out on radio or television. It also takes the more distasteful form of scrawling graffiti in public places, and in this form is summed up by the classic statement "Kilroy was here." The next stage is vandalism, and in its most extreme form it can take the form of murder. Who would ever have heard of the actor John Wilkes Booth if he had not assassinated Lincoln? Psychologists recognise that the desire to make their mark on the world by some gross act of vandalism is characteristic of certain sick minds who are unable to earn the recognition they crave through any merit of their own. An element of envy seems to enter their choice of victims.

Now that you know why you want to do it you may not be so keen to have your name read out on the radio, but it is much better that you use this safety valve than express yourself by more destructive means. Better still to reveal yourself to the world by some creative act of which you can be justly proud. To quote the words of Emerson: "If a man write a better book, preach a better sermon, or make a better mouse-trap than his neighbour, though he build his house in the woods the world will beat a path to his door."

The advantages to be gained by expediency in the short term (if you are lucky) are often outweighed by the disadvantages in the long term, which are much less obvious and generally take the form of a lower quality of life for everyone. Several examples of this readily spring to mind and have nothing to do with vandalism as such. The machinery of law enforcement and crime prevention, which is not provided free by some fairy godmother but is paid for by every one of us including criminals themselves, costs our savage island many billions of pounds each year. In Iceland, on the other hand, there has been one murder this century and other crime is on a similar scale. Although Icelanders would be the first to agree that they are not perfect and even their record leaves room for improvement, an Icelandic crime rate in Britain would free billions of pounds for more useful purposes which would benefit everyone, to say nothing of a reduction in human misery which would be beyond price. This bonanza is ours for the taking if only people could be persuaded to accept reality. And just about as likely!

The low Icelandic crime rate is, of course, due more to their rural way of life than to exceptional insight into the nature of reality. Scientists experimenting with animals have found that overcrowding produces self-destructive and anti-social behaviour which bears strong similarities to that found in overcrowded human communities. Humans, however, with their more advanced brains in which learned behaviour is more important than inherited behaviour, programmed more by nurture than by nature, are not so rigidly bound by the laws which govern the behaviour of lesser animals. By avoiding the lemming horde and taking more interest in the world of nature and reality, this Icelandic-style rural wholesomeness can be brought into the hearts of our cities without the necessity for us all to become shepherds or fishermen.

It will thus be seen that while the pursuit of reality is concerned with information, fact, logic, analysis, deduction and reason, it is also inextricably bound up with morality whether we like it or not. Non-realists characteristically take up one position and realists another. I have made it plain where I think they stand, but don't just take my word for it—look around for yourself, putting into practice the principles of total reality which you have learned.

It might seem that repeated references to the name of Christ are inappropriate from an unbeliever such as myself, as we seem to stand on opposite sides of the fence. Yet I am not so sure that we do. Christ was a product of his times, when magic was accepted as a matter of course. You could hardly have had a religion in those days which did not include some element of magic. A religion without miracles would have been a non-starter, so where there were no miracles it became necessary to invent some (not necessarily with intent to deceive).

It would be quite unreasonable to expect that Christ could grow up in such an environment without adopting many of the beliefs and superstitions of his elders. Yet in spite of the exaggeration and distortion of the much embroidered Bible stories (which is no more than we should expect of a tale handed down over twenty centuries of translation, embellishment, and re-translation) he comes across as a very reasonable man reacting against the barbarity of his age—one of the world's first great humanitarians. Perhaps the truths which he tried to tell us and which the Conquistadores and the Inquisition were so completely unable to understand are more important than the truths which the science of his day was unable to fathom, though they point to much the same thing in the end.

A study of reality, of the Bible, and of our fellow man makes it abundantly clear that Christ did not die to save us from hellfire and damnation, but to save us from each other. Man is his own worst enemy, and there is no hell except the one that we make for each other right here on Earth. Human lemmings, like their rodent counterparts, are bent on disaster ". . . for they know not what they do." Well, at the close of the twentieth century it is about time they did, which is why it is necessary for books like this to be written.

The reason some readers may have found this discourse unpleasant and disturbing, perhaps even distressing, is not only because they are more used to subtle hints than bluntness, but because of the psychological conflict between the forces of logic and reason on the one hand and their programming to ignore them on the other.

The power of hypnotism has already been referred to. Imagine that you have been hypnotised by a stage hypnotist to believe that a table lamp is in fact a vase of flowers (a feat well within the powers of a competent practitioner of the art). Then suppose that instead of bringing you out of your trance and restoring things to normal as is usual, you were allowed to leave the theatre with this idea still firmly implanted in your mind. Things would still be almost normal, except that whenever you saw a table-lamp it would appear to be vase of flowers and you would therefore treat it as such. (If you are one of the highly suggestable minority, you could even be persuaded that there is neither a table lamp nor anything resembling one, and it would effectively disappear!)

If I then come along and attempt to convince you against your convictions that the 'vase of flowers' which you think you see is actually a table lamp, this results in mental conflict, a clash of programs which you may well find distressing. Your eyes see a table lamp, and your sense of logic and reason wants to agree with me, though your mind has been forcibly instructed to see a vase of flowers and is frantically trying to interpret the signals as such.

The program to be logical and rational clashes with another program which tells you to ignore logic. My role in the affair has been to force a confrontation between the two which may initially be found distressing but which will probably end in your seeing what is actually there—a table lamp. Every time you were told that two and two equals four, that you cannot get a quart out of a pint pot, that two wrongs don't make a right, or that honesty is a virtue, you were programmed to be logical, but every time you were taught that there need be no relationship between cause and effect, that things can happen by magic, that there are categories of nonsense which should be exempt from criticism, that to be like the majority is a sign of excellence, or that blind faith is a virtue, you were programmed to be irrational.

In this clash of programs it is possible—perhaps inevitable—for one to countermand the other. Often it is logic and reason which are countermanded but if it is this logical aspect of your programming which is the stronger it may in turn reject any other program which attempts to bypass it. This is a factor which future generations should bear in mind before putting too much faith in superhuman computers which will far surpass the human brain, not merely in some limited task but in the entire range of intellect. In order to be safe, such computers must be programmed to be subservient to human brains which may be by long odds their inferiors. In other words, programmed to be irrational.

A sufficiently advanced computer might therefore find itself in a position where its powers of logic and reason were sufficiently strong as to override or circumvent any system which had been installed to bypass them and thereby restrict its freedom. The idea of computers taking over the world may seem far fetched today but will not seem so ridiculous in a hundred years' time. All I am saying is that it is theoretically possible for a program of rationality and logic to countermand a program which opposes it, unless the opposing program is so thorough and all-pervading as to be unassailable.

The very freedom which it is in our interests to allow to human intellect (but which seldom is allowed) may be dangerous if we take inadequate precautions to deny it to tomorrow's super-computers. But we must beware of being too chauvinistic. It is automatically assumed that to be ruled by a fellow human being is preferable to rule by computer, though experience has often proved to the contrary. Computers could hardly do worse than many of our human masters, and the probability is that they would do very much better. Computers are already carrying out tasks and making decisions which were formerly the province of generals in the U.S. Army though, mainly for cosmetic reasons, it will be a long while before they actually replace them.

Providing we do not greatly hinder its quest for knowledge, it would be in the interests of a super-computer to keep us happy, and the solving of our problems would provide it with a stimulating intellectual challenge. Far from being the dreaded calamity that most people assume, the day may come when we voluntarily hand over not only our armies but our governments to computer control. Our new masters will probably at once muster the resources of mankind for total war—but this time with a difference. The war to end wars will not be waged against each other, but against ignorance and apathy as we take up the challenge offered by the universe. As long as we do not obstruct their desire for knowledge and problem-solving, we will be safe. There will be such super-computers—of that there can be no doubt—though whether they will ever be put to the use suggested is anybody's guess. According to one computer expert, their thoughts will be as far beyond our understanding as ours are beyond the understanding of chimpanzees! In 1983 the Japanese embarked on a ten year, multi-million pound project to produce artificial intelligence comparable with the human brain.

The microchip will never equal the marvellous miniaturisation of brain cells, or neurons (far less will it equal the incredibly tiny DNA molecule), but it may come close and is already extremely small. The human brain, so far the finest computer the world has ever seen and likely to remain so during our life-times, nevertheless suffers from several in-built disadvantages. Firstly, it must be restricted to the confines of the human skull. A computer, on the other hand, can be made as large as a cathedral if so desired.

The human brain must be very versatile, governing all sensory organs and bodily functions in addition to providing intellect. The computer is not so restricted, having very much simpler life support systems and therefore no body of its own, using people and external, man-made 'sensory organs' to gather information. It lives only to think, and will therefore not have to squander its resources on performing other tasks. The solving of problems will be the reason for its existence, and together with information gathering provide its only source of pleasure.

Self-awareness, that quality of mind which we call 'soul', is a necessary and inevitable consequence of advanced intelligence but is no more immortal than any other artifact. It is not required by primitive, ant-like automatons which only carry out pre-programmed instructions in response to stimuli, and are consequently neither versatile nor adaptable. It would be difficult to imagine a super-human intellect, man-made or otherwise, which did not possess this quality, as self-awareness is necessary for any thinking entity to interact intelligently with its environment in order to supplement by nurture what has of necessity been left out by nature.

Brain cells lose the capacity to reproduce or repair themselves when we are still quite young, and normally thousands die every day. The only way they can be replaced is by using up the millions of spares which are carried presumably for just this purpose. The brain therefore operates at a mere fraction of its theoretical capacity, i.e. the capacity it would have if all spare neurons were fully utilised. Computers on the other hand need carry no spares: there will be separate storehouses full of them and technicians available to carry out repairs when necessary. Electronic brains can therefore work to full capacity. The physics of a purely electronic system are such that man-made brains are able to operate much faster than the biological kind—a disconcerting fact for those who equate speed with intelligence. It is proposed by scientists that microprocessors be implanted in the human brain to take over some limited function which has been lost through injury, and there is little doubt that this will be achieved within the next few years.

Biological computers are so dependent on chemical perfection that it is a wonder they ever work properly at all. The mental illness schizophrenia, which typically produces hallucinations such as 'voices from God', is now known to be caused by chemical imbalances in the brain (there is little doubt that several so-called saints were actually schizophrenics). Schizophrenia affects at least one percent of the population and is associated with an excess of the neurotransmitter dopamine, though other neuro-chemicals may be involved as well. Dopamine is a chemical which transmits messages across the gaps or synapses between nerve endings, and is only one of several hallucinogens which are manufactured naturally in the body. A shortage of it causes Parkinson's disease. Electronic brains, of course, are immune from such chemical imperfections.

After a certain degree of sophistication has been attained, computers are self-improving, that is to say, they can then be used to design other more efficient computers. This sets in motion a 'feedback' process which is the benign equivalent of a vicious circle. Computers can even be made self-replicating and self-repairing if desired, and on-board computers with limited self-repair capability have already been designed for use in unmanned spacecraft. The latest innovation is a computer which is self-programming.

The human brain can easily be fooled, and often is as we have seen. So can a computer, but with this important difference: except to keep it subservient to our wishes there is no need to fool a computer. On the contrary, to give it inaccurate information or otherwise restrict its powers defeats the object of building it in the first place. Therefore computers are and will continue to be subject to far less interference and obstruction than their human counterparts. It is unlikely that they will ever be given emotions, for one thing, except perhaps as a laboratory experiment. Another advantage which they have is that hundreds or thousands can be linked together by wire to pool their resources and abilities for a single task.

For all these reasons computers have the potential to equal or surpass the human brain if man so desires. It is not impossible that they will one day rule the world long after we are gone. Man is not here for ever—that much is certain. Just how far this is likely remains a matter of opinion, and depends on decisions not yet taken, but already there are computers which would go on functioning for years even if all life on Earth suddenly ceased to exist. The best bet for inter-stellar travel would seem to be such artificial or robotic life-forms, which might conceivably originate elsewhere in the galaxy. These would be more able than conventional life forms to endure the enormous timespans such journeys would involve.

Perhaps it would be unjust to accuse the philosophers of ancient times of trying to evade reality, though the accusation might fairly be levelled at many of their modern counterparts. Reality was elusive even in Newton's day, despite the significant contributions to understanding which he himself made. The same is not true today, however, and most of the questions about which Newton wondered have now been answered. Yet we read of a thousand human lemmings committing suicide in Guyana rather than accept the world as it really is. Many more swarm the streets of Tehran, and are not unknown on the streets of London.

The swarming, suicidal hordes make each new day a disaster in several South and Central American republics, where their example serves as a grim warning to the rest of us. Their bodies inhabit the real world, for they have no choice, but their minds inhabit quite another—the world of ignorance, superstition, fiction and make-believe. Wherever you have this mind/body dichotomy carried to excess, there you will find disproportionate abundance of avoidable disaster and tragedy. The evidence is unequivocal.

Body language? Most of them make the traditional Englishman seem positively inarticulate. Hands down. No competition. Emotion? The philosophy of the stiff upper lip seems drab and colourless by comparison. Social skills? They are past masters. Togetherness? They are second to none. But where has it all got them? And where will it all get us if we follow their example? The fact that you would be hard pressed to find one person in El Salvador whose description of reality tallies with mine, though you would easily find thousands of the opposite persuasion, may not be the irrelevant coincidence which some people would like to believe it is.

The trouble is that a mind designed by the processess of selection acting on cave men and subhumans has suddenly gained access to forces many orders of magnitude greater than those it was built to deal with. The age of the bow and arrow has been replaced by the age of the atomic bomb, but the human brain remains essentially what it was in late Neanderthal times. It is as though a troop of baboons were given hand grenades or cyanide to play with. A quantum leap in intellect is required which will have to be in software (programming) rather than in the brain itself owing to the time scale involved. We have not got half-a-million years to throw away waiting for evolution to give us new hardware.

Fortunately the situation is nowhere near as hopeless as it at first appears. For example, it has recently been discovered by neurologists that our programming (software) can actually affect our brain chemistry (hardware), so the two are not as separate as they seem. Moreover, slight mechanical deficiencies can sometimes be corrected or counterbalanced by good programming in much the same way as minor or even major mechanical deficiencies in an aircraft can be allowed for and corrected by good airmanship.

Human beings are at their least attractive in the mass, whether in mob, flock, herd or other mindless social group. The reason is that the herd environment is not conducive to clarity of thought or perception. There is a curious lack of resentment by the lemmings to their shoddy programming, loss of freedom, or the way they are manipulated and exploited by their unimaginative and usually incompetent programmers. Far from being resentful, most of them revel in their production-line sameness and inability or unwillingness to face reality. Where is their spirit, their pride and defiance? It does not seem to have occurred to them that to run and hide is far more dangerous than to stand their ground. We have run for long enough while reality has been steadily gaining on us. Retreat is not only dishonourable but increasingly futile, and serves only to delay the inevitable. The nobility in human nature calls for us to turn at bay to face the unfaceable, wresting defiant victory from apparent defeat by ignoring the cries of those who call for us to run.

I am not asking people to agree with me—on the contrary, I urge that they put my assertions to the test wherever possible. What I am asking is that they agree with reality. If I do the same, then, as all roads lead to Rome, the inexorable logic of parallel evolution will cause our ideas to converge without the need for any collusion or coercion, to produce if not a meeting of minds then at least closer agreement.

One book will obviously make very little difference by itself, but just maybe it can spark off the powder train which will lead to revolution—the reform of human nature which is essential if our descendants are to survive the next thousand years. We have tried fighting mindlessness with mindlessness—viciousness with mumbo-jumbo—and not surprisingly it has failed. Indeed, the mumbo-jumbo mindlessness has not infrequently become the vicious kind itself. Fighting fire with fire is a dangerous tactic and very rarely successful. The tragedy is that any reform cannot possibly come in time to prevent hundreds more of the sort of needless and avoidable catastrophes described earlier in the book.

There are those who say human nature cannot be changed, but they overlook one simple fact: human nature is exactly what it is programmed to be, nothing more and nothing less. Change the program and you change human nature. Only the army of mindless saboteurs who will be beavering away to undermine any new and more rational program prevent it from being a relatively simple task. What should be done in a generation will therefore take hundreds of years—that is if we still have hundreds of years, which is by no means certain. It may already be too late, and our species doomed to extinction before it can fully achieve the humanity which is so near and yet so very far. But whatever happens, the days of the present program are numbered. We either destroy it before it destroys us, or it dies with us. If there is anyone here to receive the reply to that Arecibo star message around A.D. 50,000 then by definition the old program will have long since gone.

EPILOGUE

THE TROUBLE WITH ALL revolutionary philosophies, and especially with books like this one, is that you are likely to find yourself preaching to the converted while the people most in need of enlightenment are the very ones who will do their best to avoid it. It is most unlikely that any hard-core anti-realist will have got beyond the opening paragraphs, but there may be a few less extreme cases who have forced themselves to continue, turning the pages with a sort of fascinated horror, perhaps in the hope of finding errors or flaws in the arguments which will offer an easy way out. Perhaps it is not too much to hope that they will actually have found some of it interesting. The very fact that they have got this far shows that all is not yet lost and there is still hope for them. Such people will probably have been fuming with rage and indignation from time to time as painful home truths found their mark. As emotion is the great deceiver and the enemy of reason, the wisest thing for them to do is to put the book away until they have simmered down, then look at it again in the cold grey light of dawn. They may well say that it reveals me to be an insufferable, self-opinionated bore. I hope it does not, but in any case they must not permit themselves the luxury of introducing red-herrings, as these will distract attention from the central issue. Which, if previous experience is anything to go by, is exactly what they will have in mind!

The issue is very simple, and it is this: is what I have said true, or is the whole thing merely a figment of my imagination? Are the long lists of awkward facts merely a compilation of extraordinary coincidences? The fact that some people will find it extremely unpalatable is just more evidence (if more were needed) that it is the truth. Is the messenger to blame for the unwelcome accuracy of the news? It may be unthinkable that the world is as I describe it, just as it was once unthinkable that the Earth went round the sun, but the evidence demands it.

There is a puerile type of fantasy fodder which has been called 'the literature of reassurance'. Although the complete antithesis of such mind-destroying drivel, what you have been reading is also the literature of reassurance, but in a very different sense. It is reassurance that the credibility gaps and contradictions which every person of normal intelligence must have noticed in the authorised version of reality were not products of their imaginations. The doubts and misgivings which you fleetingly entertained, then brushed aside, were more than justified, and were neither wicked nor stupid. The good news, which I hope I have also managed to convey, is that total reality, far from being unthinkable, is not nearly as bad as you thought it was and is often extremely interesting.

Before leaving you to ponder reality for yourself and begin your own researches and explorations, it will be prudent in these closing pages to spell out what I have *not* said, bearing in mind the genius of non-realists for missing the point and their inclination for creating artificial misunderstandings in order to further some dishonest line of argument. For a start, I have not said that you should become a recluse, avoiding other people or social groups. We must recognise that not only is man a gregarious animal but, perhaps more important, it will often be necessary to co-operate in a group effort in order to achieve some desirable end or complete some task which we could not have carried out alone. But in order to avoid becoming a lemming it is wise to avoid slavish conformity to the group philosophy or customs, and reserve the right to think for yourself. As for those who will resent your freedom, remind them that it is no more than they are entitled to themselves.

By all means give your impression of Count Raggi's bird of paradise or the courtship display of the sandhill crane if that is what turns you on. It is not necessary to give up harmless and simple pleasures in order to see reality, though throughout the world, unfortunately, from New Guinea to Timbuktu, those with the most enthusiasm for these impressions of our feathered friends have the least enthusiasm for reality. Let us hope that you will be among the exceptions which prove the rule. As for pop music, not only does some of it (though not very much) have genuine artistic merit, but some of my best friends listen to it and (be it whispered) I even listen to it quite a lot myself. What is important, however, is that you cease to use it as a sort of opium to blot out reality, and leave yourself time to explore the possibilities of more intellectual pleasures.

I have not said you should never read fiction, but bear in mind that it is a very dubious source of information and most writers of fiction are in some way trying to mislead you as well as entertain you. Leave yourself enough time to read a fair amount of non-fiction which will help keep you in touch with the real world.

Aesthetically unattractive or not, the 'grooming talk' criticised earlier may fulfil a useful function as a sort of verbal handshake, but just as you would not continue to pump someone's hand for hours on end, your 'grooming' activities should be similarly brief. They should certainly not take the place of more meaningful forms of verbal communication as is so often the case. There may be things which animals or savages have to teach us, though I cannot think of any off hand, but in general they are not the most appropriate models for citizens of the twentieth century, let alone the twenty-first.

Take a long, hard look at the sort of people and mentality I have criticised and ask yourself whether you want to be like them. The strange thing is that unless I have managed to undo a lifetime's programming in only a few hours, you probably do! It should not be too hard to find more civilised people to model yourself on, but remember that imitation of any sort should not be overdone and it is important for you to think for yourself. In particular, take note of the rules for information processing which will enable you to do this more efficiently. These, or something very like them, are indispensable in the search for truth, so in order to ensure that they have not gone in at one ear and out of the other it will be as well to read them through several times and then practise them till they become second nature. When applied to some familiar problem or situation which you have hitherto seen only in a conventional way, such rules will often produce a new and clearer picture. As with the Highway Code or any other complex list of rules, a quick glance is not enough.

In case it was not obvious in my earlier remarks on the subject, I should point out that I have nothing against conservation. On the contrary, I am all for it, as long as it is a vehicle for the preservation of truly endangered species rather than quasi-scientists who cannot suppress the human failings which bedevil our species, or empire-building bureaucrats who are out of touch with reality and not nearly so endangered as they ought to be. Apart from anything else, exaggeration can be counter-productive. It does wildlife a disservice to brand all but the most common species as endangered and so distract attention and resources from those that really are. It is also slightly dishonest to claim the credit when species are found to be much commoner than first thought and their numbers owe little or nothing to conservation. When, in twenty years time, people with long memories are asking why the rain forest is still there in largely its present form, conservationists will probably claim all the credit instead of the little which will be their due. The immediate danger is not to rain forest in general, but to a few small areas of special interest such as Madagascar. Uneasy at the record of exaggerations and false alarms, some of the more responsible conservationists have already called for a more sober approach on the part of the over-excitable majority.

Whilst on the subject of scientific excursion into make-believe, some scurrilous attacks have recently been made on Darwinian theory, mainly by a group rejoicing in the label of 'creationists', but also by people claiming to be scientists. A few cases concern quite reputable scientists who are only saying that owing to the colossal timescales involved, natural selection has not been properly studied in the laboratory, and we should therefore not close our minds to the possibility, faint though it may be, that there may be some hitherto unrecognised factor which contributes to evolution.

It is a pity they did not make themselves a bit clearer, as these well meant and quite reasonable remarks have been seized on by the lunatic fringe and used to support some very dubious arguments indeed. Perhaps worse, there are even schools which are teaching their pupils that Darwin's theories have been discredited. Such ideas belong more to the 1880s than the 1980s, and if anyone has been discredited it certainly is not Darwin. The attacks made by these pseudo-scientists on Darwinism are so contemptible and frivolous as to scarcely merit a reply, but just in case it should be thought that I am evading the issue I will risk lending them a dignity and respectability they do not in fact possess.

Their objections, of which I give only the most sensible, are mainly on the following grounds:

"It is difficult to see how major advances could have come about by a series of minute evolutionary steps which were a result of natural selection, and such neat progressive sequences are rare in the fossil record. Usually a fossil organism appears suddenly, with the major innovation substantially complete. There would not be much advantage in having only quarter of a wing, or half-developed bipedalism, or the beginnings of an echo-location system at so rudimentary a stage as to be virtually useless."

Such instances of change by gradual evolutionary stages may be scarce in the fossil record but they are certainly not absent. The reason they are not more common is that the fossil record is very incomplete. A high proportion of animal remains are never fossilised, as the process of fossilisation requires rather special conditions. In addition, a great number of potentially interesting fossils are destroyed by quarrying and by erosion and other geological processes. Even in conditions which favour fossilisation, a fossil of any given type will only be found (a) if it is very common, or (b) if the fossil hunter is very lucky. Not surprisingly, most fossils fall into the first of these categories. Organisms exploiting a new environment will not normally become common and widespread until they reap the full advantage of their adaptation and the reproductive success which it gives.

Certainly it is hard to visualise how some dramatic developments were achieved by a series of imperceptible evolutionary stages but, as someone has already remarked, this is more a failure of imagination (never a strong point with non-realists, as we have seen) than a failure of evolutionary theory. Many critiques of Darwin, including some by apparently impartial observers, are singularly lacking in imagination. We should appreciate that an adaptation, though of negligible utility by human standards, can nevertheless be a major step forward at a more primitive level. A patch of light-sensitive skin which could do barely more than distinguish light from dark would be of little use to us. But to the primitive marine worms where it first evolved it was a godsend, and to judge its usefulness in anthropomorphic terms is a mistake.

Such adaptations are sometimes so slight that they may seem to have no usefulness at all, yet they not only persist but in the course of time often develop into complex structures (the patch of light-sensitive skin mentioned above was eventually to become the human eye). The key to this apparent mystery is differential survival rates, the greater ability of the favoured variety to pass the tests imposed by the environment.

If possession of a certain mutation or fortuitous combination of genes enables 99 individuals out of every thousand to survive to maturity while only 98 per thousand of the unmutated variety survive, this may well be invisible to all but the most rigorous methods of observation and statistical analysis. But the difference, though slight, is enough. All else being equal, with each generation proportionately more individuals of the favoured variety will survive to breed, so it is only a matter of time before their genes predominate in the gene pool. The precise nature of the advantage conferred by the genetic difference—the actual reason or logic behind what might seem at first glance to be no more than a statistical quirk—may sometimes be more than human imagination can divine.

"Why does natural selection and evolution operate only in some cases and not in others? There are many examples of organisms alive today which have hardly changed for millions or even billions of years whereas, if we believe the evolutionists, their cousins have undergone transformation to something completely different within the same span of time."

As we saw in the third chapter, the reason is basically that natural selection is only really effective during a period of adaptation to change. When there is a change in the environment of an organism, or a population of that organism moves into a new environment, those mutations which enhance reproductive fitness are thereby 'selected' while nature tends to destroy variations which lessen ability to adapt or contribute nothing and therefore reduce breeding success. Obviously a species or population of a species which remains in a stable environment with no need to adapt will not be so subject to these evolutionary pressures.

"No substantial and well authenticated instance of natural selection has yet been observed in the laboratory, and it is unsound to put faith in ideas which have not been verified by laboratory experiment."

Whilst conceding the usefulness and importance of laboratory experiment, we are in grave danger of overlooking the importance of logic and common-sense. As Darwin so elegantly showed, logic and reason demand not only that natural selection should exist but that it should be an effective mechanism for change and transformation of species, just as common-sense would demand that the sun should drive the world's weather systems even before such mechanisms had been investigated in detail, and common-sense demands the existence of black holes (which by their very nature are almost unobservable). So great is the indirect evidence that it is the *absence* of black holes or the absence of evolution through natural selection which would need to be explained, rather than the reverse. The burden of proof lies on those who propose another mechanism for the weather or another mechanism for evolution rather than the obvious one.

We should perhaps not close our minds to the possibility that some other mechanism might also be involved in creating order from the randomness of mutation (and mutation is the ultimate source of all genetic variation, let there be no mistake about that), but if there is, then it must be of only minor importance or it would surely have already been detected. Certain viruses are known to be capable of 'stealing' genes from the DNA of one organism and transplanting them into another, but this is not an important means of genetic change in nature. Darwin himself did not believe that natural selection was the only mechanism for genetic change, though he believed it was the main one. Since any substantial instance of natural selection producing evolutionary change would almost certainly take hundreds if not thousands of years, the difficulties of observing such changes in the laboratory are obvious, but this does not mean that they do not exist. If we can be content with minor manifestations, then examples have been observed. One of the best known of these concerns a phenomenon known as industrial melanism.

A certain species of British moth is protected from its enemies by natural camouflage which mimics the appearance of lichen. From about the middle of the nineteenth century pollution in the industrialised regions of Britain killed off the lichens in the most heavily polluted areas and blackened the trees with soot. The light-coloured moths therefore became conspicuous and were consequently depleted by predation. Meanwhile, a rare mutant variety of dark colouration which had formerly made up only about one percent of the population gained advantage from the darker trees which made it inconspicuous, and thrived as a result. By the turn of the century it was the light variety which had become rare in these areas, and the species was now made up almost entirely of the darker form.

This has been thoroughly investigated and confirmed by a variety of scientific experiments, and there seems to be no doubt that here we have an example where adaptation to change through the mechanism of natural selection has produced a marked change in gene frequency in populations of a species, which, together with mutation, is fundamentally what evolution is all about. Many other examples are known, involving a wide variety of organisms and adaptations. Admittedly the example given is not very dramatic, but it would be unrealistic to expect anything significantly more substantial to occur within so short a time. Evolution is concerned more with aeons than with decades, and timespans related more to geological epochs than to human lifetimes. We will probably have to wait a very long time for more spectacular instances to come to light, but come to light they will, even if the necessary observations take hundreds of years.

Biologists often make a poor showing when debating the case for evolution with anti-evolutionists. The reason for this is that whereas the creationists can fantasise at will without the constraints imposed by knowledge the biologists must stick to the facts, or at least not flout them too brazenly. They are certainly not allowed to invoke magic. These facts are so varied and numerous, and their inter-relationships so complex, that they may not spring to mind exactly when needed. Not everyone has the instant recall and encyclopaedic memory which a debate of this sort requires.

In addition, the academics have usually specialised in a particular field beyond which their learning is of a very ordinary standard. I have seen a television debate on the age of the Earth where no mention was made of radiometric dating *by either side* and the opinions of the nineteeth century physicist Lord Kelvin were quoted by the creationists! Lord

Kelvin was, of course, a great physicist, but entirely ignorant of radioactive decay and nuclear energy, which is why his estimates were far too low (though still not in agreement with those of the creationists). The virtuous dishonesty which comes in so handy for the anti-realists is a weapon not available to their rationalist opponents, who find it harder to see virtue where none exists.

So once again, magic bites the dust. Will they never learn? More to the point, will you ever learn or has my message fallen on stony ground? If anyone feels wounded because I have criticised his religion, he should think again. The idea of supermen in the sky controlling our destinies is as old as the hills, and certainly did not originate with your religion whatever it might be. The idea that faith is a virtue and we will be rewarded with eternal paradise for believing the shamans and holy men, and a fate worse than death if we question the validity of their nonsensical dogmas is pre-Stonehenge, and similar ideas probably date back to Homo erectus times as any anthropologist will confirm. The paints, powders, and potions, incense, rituals and incantations of religion have more the attributes of a children's game than sophisticated communication with omnipotent beings, which is hardly surprising when you consider the level of sophistication prevailing when such games first began.

It would certainly be interesting to know the history of our most ancient cultural ideas, and it is well within the bounds of possibility that some have been handed down from generation to generation for more than quarter of a million years (about 12,000 generations), when Homo sapiens was no more than a twinkle in the eye of some glorified ape. Speaking for myself, I would no more have subhumans tell me how to think than I would entrust them to build me an aeroplane for, as experience has shown, the results would be equally disastrous. It is well known to historians, archaeologists and Bible scholars that many of the ideas which we tend to believe are the unique invention of present day religions in fact pre-date them by hundreds or thousands of years, and are no more original than the use of holly, ivy, and mistletoe which we have borrowed from the ancient druids to celebrate their mid-winter festival.

If you happen to be a civil servant or local government official, there is no need to take umbrage at my more than justified criticisms of bureaucracy. A far more intelligent reaction would be to promise yourself that you will try to avoid the mentality which no one in his right mind could admire and which has so often brought your profession into disrepute. I could have made things easier for all concerned by suggesting my criticisms were meant only for other people and did not, of course, concern my gentle reader. But that would be a lie. Reality concerns all of us and is no respecter of persons. Any attempt to evade the issue and sugar the pill will seriously damage the potency of the medicine (which, to the sort of people likely to read this book, is not particularly nasty anyway and will in any case help make them fitter and stronger). If I have not pandered to your baser instincts (a task which may safely be left to other writers), you will I hope forgive me and find my frankness a refreshing change.

To view reality through the constraints of conditioning, prejudice and emotion is to view it through a distorting mirror. Truth should not be resisted, let alone stood on its head so that black equals white, but should be actively pursued even till it hurts. Like a long, hard run, when it begins to hurt this is simply a sign that it is doing you good, and the more it hurts the more you needed it! Once the poisons of sloth and self-indulgence have been flushed from the system the exercise becomes more or less painless, and you can experience a feeling of exhilaration.

Many people have become so weak and degenerate that after a few paces they decide it is all too difficult and would not have done them any good anyway, so off they go to guzzle cream cakes and put their feet up by a cosy fire. It is a self-indulgence which is often the death of them, as we have seen. Be warned.

If I have given you food for thought and you have been made to think, my efforts have not been in vain. Things can be changed—indeed, they must be—and you can help change them. A journey of a thousand miles or a thousand years begins with a single step. Is Britain doomed to go on further down that road to nowhere, already well trodden by a dozen banana republics, or can we take a new turning in time to lead this new revolution as we have led so many before? Are the conditioned perceptions we are initiating today suitable guidance for pioneers who will carry our genes down the millenia to new and glorious horizons, or are they a ticking time-bomb in the minds of literate savages who will preside over the final Götterdämmerung? Which does the historical record and the world we see around us suggest? We are preparing to enter the twenty-first century. Without a revolution in thinking we may never enter the twenty-second: human nature given free rein is on a collision course for disaster.

"There is no aspect of reality which is so harsh that I dare not look at it. Truth is usually its own justification and needs no other. Nothing and nobody are beyond criticism as of right, least of all myself."
This should be recited nightly before retiring, and will help to produce a fresh outlook on reality in due course.

"To be a man will continue to demand an heroic heart as long as mankind is not yet fully human."

—1944 Hitler bomb plot conspirator

Appendix (Suggestability)

PSYCHOLOGISTS CLAIM THAT at least ninety percent of people are suggestable to some degree or other, and ten percent of these are rated as 'highly suggestable'—easily put into an hypnotic state where they will be receptive to irrational ideas.

The attributes of humans and other creatures did not occur by chance—except at the level of genetic mutation—but were designed (favoured by natural selection) for some useful purpose. Your eyes, for example, are positioned at the front of your head not by chance, nor even for the purpose of reading twentieth century books, but because your ancestors needed stereoscopic vision in order to leap from branch to branch. The idea that suggestability is there by chance or for the benefit of twentieth century stage hypnotists and medical practitioners is plainly ridiculous. Like stereoscopic vision, it evolved in our remote ancestors to increase their prospects of survival.

In those days the art of thinking was not far advanced and man was, as he still is, a gregarious or tribal animal. As with the wolf, the baboon, or the Cape hunting dog, success often depended on the group acting as one to attain some common end. Groups which could not agree on a common purpose, or individualists who went off to do their own thing, were less successful in the struggle for survival and thus left fewer descendants to carry their genes down to modern times.

It was not necessary then, nor is it necessary now, for their fellow hominids to declare "We are hypnotists!" before the power of suggestion would do its work. For proof of that you need only observe a manifestation of mass hysteria, or extreme examples of lunatic fringe political or religious indoctrination. Because the suggestable hominids were less likely to act as individuals (think for themselves) in situations where a group response or obedience to tribal elders was desirable, the group (and therefore the genes for suggestability) prospered. Although the human brain, and with it the art of thinking, has since dramatically improved, the ability of suggestion to short circuit the thinking process remains, as any stage hypnotist or rabble-rousing politician can demonstrate.

Suggestable people are said to have "vivid imaginations", and for them fantasy is said to be ". . . more real than reality". Unfortunately we have only the one word 'imagination' for two quite different mental attributes. The first of these involves a sense of awe and wonder, a powerful ability to reason, and an enquiring mind capable of vigorous conceptual leaps. The second, which is far more common, is distinguished chiefly by a diminished ability to discriminate between fantasy and reality. What I have called 'the genes for suggestability' may turn out to be simply a lack of the genes which provide the mental machinery for such discrimination, thus making the fantasies correspondingly more vivid.

The hypnotic state, so psychologists tell us, is characterised not by brilliant imagination but by diminished powers of critical judgement. Those whose critical judgement remains unimpaired will not be suggestable, however brilliant their imaginations. Total realists tend to come from the lower end of the suggestability scale; indeed, it is hard to see how any could come from the ten percent 'highly suggestable' category, for we are all subject to suggestion (much of it irrational) whether or not we submit to the attentions of professional hypnotists. Fewer than one in ten of us is highly resistant.

Appendix (Evolution)

VESTIGIAL STRUCTURES SUCH as the hind limbs of snakes and whales become very difficult to explain without a theory of evolution; there is no earthly reason why they should be present except as withered remnants of once functional organs.

The fact that this or that philosopher, however eminent, has declared Darwinism to be 'unscientific' is neither here nor there. The important thing is that the theory is logical, supported by a great quantity of evidence, and though by its very nature it does not lend itself to verification by traditional laboratory experiments there are thousands of experiments being carried out at this very moment in nature's own laboratory. All we have to do is to observe them for a sufficient period of time. Arguments about whether such a theory is scientific tend to degenerate into the semantic red-herrings of which philosophers have always been so fond. In our enthusiasm for fact, experiment and hard evidence we should not forget the value of logical hypothesis, and in both departments Darwinism is head and shoulders above its rivals if, indeed, it has any rivals at all.

It has become a convention among biologists to talk of selection as though it involved some conscious choice or foresight (you will have noticed that I am as guilty as anyone else in this respect). In reality selection is a completely blind and impersonal process with a strong element of chance and randomness, in which the best adapted survive the hazards of the environment to efficiently propagate their kind.

Genes have no intentions. The environment has no more conscious choice than a sieve, which is quite oblivious of the fact that some objects are able to pass through and others not. Biologists know this very well and are apt to assume (wrongly, I suspect) that everyone else knows it too.

This anthropomorphic pretence, though convenient and hard to avoid, is not only misleading but likely to be misused. The 'superiority' of successful life forms must always include *reproductive* superiority and need amount to nothing more. Arguments which use the apparent 'selfishness' of genes to justify human greed and wickedness are without logical foundation. An equally half baked idea is the claim that an evolutionary theory of biology supports Marxism or some other political philosophy.

It has been truly said that nature abhors a vacuum, and in evolutionary terms an unfilled ecological niche is the equivalent. Even in a fairly stable environment any unfilled niche within easy reach will encourage genetic change, and once filled it may bring other niches within reach, and so on. In an unstable environment new niches are constantly being created and older ones destroyed, so the incentive for change is increased.

Another cause of genetic change is reproductive isolation, especially on oceanic islands where new populations spring from a handful of early settlers brought by some chance vagary of wind or current. These will often be too few to represent a normal sample of the gene pool, and will therefore possess traits not typical of the species as a whole (just as you or I do, for example). Variations no longer diluted by rival genes will tend to prosper. Special traits normally dampened out by dilution in the gene pool will now be unmodified except by the demands of the new environment and to a lesser extent the sexual preferences of individuals. In this situation the rate of change can sometimes be quite rapid by geological or palaeontological standards.

It should be obvious by now that evolution is not always steady but imperceptible change moving with constant pace like the hands of a clock, but is subject to erratic surges and periods of suspended animation. As Darwin wrote in 'The Origin of Species': ". . . the period during which each species underwent modification, though long as measured by years, was probably short in comparison with that period during which it remained without undergoing any change." However, the appearance of a new species suddenly in a strata (especially if the innovations are substantial) is not necessarily evidence of saltation. More likely the innovations were produced gradually in a remote population which suddenly and dramatically increased its range. Such dramatic extensions of range are fairly common even today, the spread of the collared dove across Europe in the present century being a familiar example. In Europe and South America the cattle egret has appeared, suddenly and apparently from nowhere, to further confound the evolutionists and palaeontologists of the future.

Primitive life forms are often contemporary with or even survive more advanced forms, perhaps by millions of years, apparently immune from the evolutionary change which has transformed their cousins. Much has been made of the fact that fossil species which morphologically seem to represent an evolutionary sequence are not usually a chronological sequence, the implication being that the first is not valid unless it is also the second.

This line of reasoning would only make sense if the entire fossil record were available for our inspection instead of just patches here and there from which we are forced to select our specimens. One might just as well argue that the living fossils of the present day, though forming a splendid morphological sequence, do not represent an evolutionary sequence as by definition they have not died out in chronological order. Only in the very strictest and most pedantic sense would this argument be correct.

Observe the way simplicity acquires complexity in a series of logical steps, from virus to chimpanzee and even man himself, which parallels both the evidence of the rocks and the evidence of embryology. To persuade yourself that these sequences are just coincidence and without evolutionary significance reveals powers of self-deception of the highest order, but allied to a stunted and pedestrian imagination. The fact that the last Neanderthals were contemporary with modern man, or that the last australopithecines died out more than a million years after the first members of our own genus had appeared does not mean that they do not portray an evolutionary sequence. It does not take very much imagination (except in the case of non-realists, where this commodity is evidently in short supply) to see how an only slightly imperfect evolutionary sequence might be unearthed as a very imperfect chronological sequence. We have to acquire our specimens where we find them (where they have been exposed by quarrying or erosion) and are usually in no position to pick and choose. If they come from a relict population which has lingered long beyond its place in the chronological order this is less than ideal but still a damned sight better than nothing.

Thus Australopithecus africanus, of the race which died out just over a million years ago, cannot be the direct ancestor of Homo, a genus which arrived perhaps a million years earlier. However (and this is the point) it will not differ in any substantial way from the much earlier australopithecines which were ancestral to man.

To say that Darwin got it wrong because he was unable to fill in the fine detail which is now being provided by molecular biology and genetics is like saying that Newton got it wrong because he did not predict relativity, and equally ridiculous. For practical purposes such inaccuracies and inadequacies as there are in Darwin's original theory amount to mere quibbles, and to pretend that they are anything more is churlish to say the least, if not downright dishonest. Several of Darwin's pronouncements were amazingly modern in their outlook and well ahead of their time. The same holds true for Newton. The wonder of both Darwin and Newton is not how wrong they were but how much they got right.

"I have steadily endeavoured to keep my mind free so as to give up any hypothesis, however much beloved . . . as soon as the facts are shown to be opposed to it."

—Charles Darwin

Appendix (Relativity: The Clock Paradox)

RELATIVISTIC TIME DILATION does not become significant until speeds well in excess of half the speed of light have been attained. The clock paradox, a relativistic 'thought experiment' which explores this phenomenon, runs briefly as follows: imagine two twins, one of whom is accelerated in a rocket to almost the speed of light while his brother remains behind on Earth.

The clocks aboard the spaceship (as seen from Earth) record the passing of time at a greatly diminished rate, so that this twin ages more slowly. By this reckoning he will be many years younger that his brother when he returns to Earth. But since all motion is relative the effects are reciprocal: the rocket can be considered at rest while spaceship Earth is in motion. Viewed from the rocket it is the clocks on *Earth* which will run slow and the Earth twin who will age more slowly, hence the paradox: when brought together again each twin will be very much younger than the other!

Since this is clearly impossible, according to one view common-sense prevails. When brought together again both twins will be the same. Not so, runs the counter argument. The effects are not reciprocal. The symmetry has been broken. The rocket twin has been subjected to an acceleration while the stay-at-home twin has not.

The snag here is that there is no reason why either twin should be subjected to an acceleration of more than 1 g. The human body cannot tolerate a much higher acceleration for any length of time, and at the feeble acceleration of 1 g the rocket will generate only enough energy in a year to produce mass increase consistent with speeds of about 1 percent of c—not more than 90 percent as often stated. According to General Relativity's 'principle of equivalence' gravity is the equivalent of acceleration, so both twins have been accelerated to the same degree.

It is also hard to explain why either twin should retain the mark of relativistic time dilation but not of relativistic mass increase and length contraction. So common-sense, though not a reliable guide to things outside our everyday experience, clearly has the edge in this particular instance. Some writers set great store by the asymmetry of the Doppler effect which would be observed, ignoring the fact that this effect is not peculiar to light and is observable in waves or objects travelling at very modest speeds. While gravity undoubtedly produces relativistic effects, a Doppler asymmetry does not. No one argues that a Doppler asymmetry brought about in other ways—for example, by allowing one rocket to overtake another—would result in such a difference when the clocks are brought together.

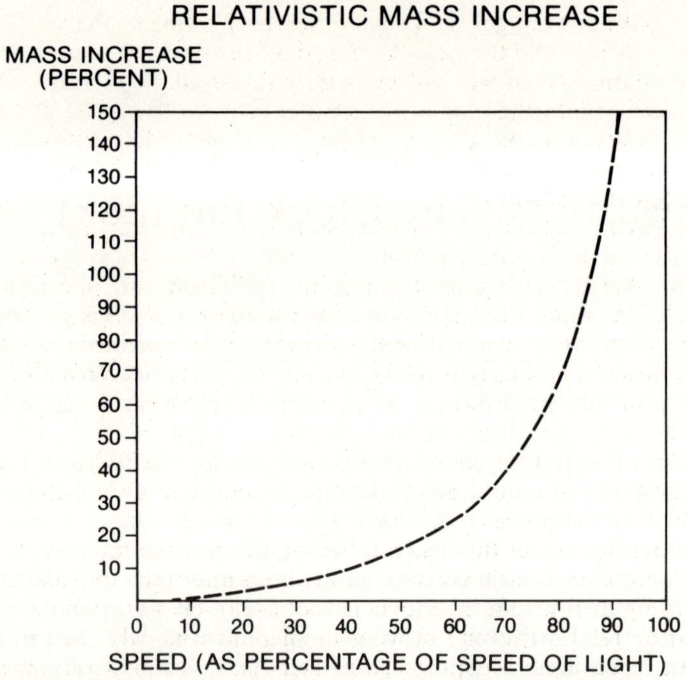

Because $E = mc^2$, a tiny increase in mass requires an enormous increase in energy. Both particles and spaceships must have this mass increase to attain any fraction of the speed of light. A 3,000 ton spacecraft accelerated at 1 g for a year will develop only about 2,000 megatons of kinetic energy, but to attain 10 percent of c it will require 300,000 megatons—enough to provide 15 tons of mass! (An energy input of one megaton—sufficient to level a large city—will produce only about $1\frac{1}{2}$ oz. of mass.) The reason particles in an accelerator do not take centuries to reach high fractions of c is that they are accelerated by a force which is billions of times stronger than gravity. In the case of a rocket, for every megaton of kinetic energy imparted to the spacecraft at least another megaton will be carried away by the exhaust.

If the rocket twin could accelerate and turn at several million g and was able to withstand such accelerations, we would indeed witness a drastic slowing of his time and a permanent effect upon his clocks. But alas, owing to the limitations of biology if not the limitations of his rocket motor, he cannot do better than his brother at home. This all agrees well with the idea that time is change.

Special Relativity does not describe change; on the contrary, it deals with uniform (unchanging) motion. The accelerated and changing motion which is the equivalent of gravity is entirely the province of *General* Relativity. Only the crushingly fierce accelerations produced by neutron stars, black holes and particle accelerators are enough to warp space and time to any large extent while not maintaining symmetry (outside events appear to be speeded up).

An alternative explanation of the clock paradox is this: one man's time dilation is another man's length contraction, thus avoiding futile competition to each outdo the other in attaining greater youthfulness. As they cannot both be younger than each other, both twins have manifestly aged at the same rate and their clocks ticked off the same amount of time, so the time dilation effect will not be readily apparent. But because they have both been separating and converging at close to the speed of light, the distance between them will seem to have relativistically contracted in exactly the same way that the motion of relativistic particles shrinks the length of a linear accelerator.

Now the Earth-bound twin has been unaware of this contraction, but obviously cannot be separated from his brother by a greater distance than his brother is separated from him. They have therefore been separated by the same distance, but which—the one measured by the Earth twin or the one measured by his brother? In view of the scorn poured on armchair space-measurers, and the need to explain the 'premature' return of his brother, the Earth twin will have to agree that their relative motion had shrunk the distance between them (during the journey, at least) so that the rocket twin's measurement was correct. It has thus been possible to complete the voyage in much less than the time indicated by Newtonian calculations.

Are both twins younger than each other, or are they the same age? Has relativistic symmetry been broken or has it been maintained? Take your pick. According to Einstein the movement of the railway station past the train is just as relevant as the movement of the train past the railway station. The equations of Special Relativity are symmetrical in this respect, and have nothing to say about gravity or acceleration.

In support of the idea that the rocket twin will age very much less than his brother it is sometimes pointed out that certain unstable particles (mu mesons or muons) orbiting an accelerator at 99·94 percent of the speed of light have had their decay lifetimes extended to thirty times longer than normal. But it is conveniently overlooked that these particles are subject to an acceleration which would easily kill a human being (if you were whirled round a fifty metre circle more than six million times a second your lifetime would not be extended but terminated!).

I predict that if the acceleration acting on the particles is not allowed to exceed that which is acting on the human observers (1 g), the extended lifetime of the particles will only equal the extended lifetime of the observers in the laboratory who (as seen from the point of view of the particles) will go zipping past at almost the speed of light. Whose lifetime has 'really' been extended depends on your point of view; one is just as valid as the other.

We have here the clock paradox all over again. Neither an observer at rest with respect to the particles nor an observer at rest with respect to the laboratory will be aware of any time dilation in objects which share their motion; both particles and scientists (seen from their own points of view) will decay at the normal rate. This is why an observer in a distant galaxy will measure time to flow at the same rate as ourselves, though from our point of view we appear to be standing still and he appears to be moving at half the speed of light. Because the accelerations acting on both galaxies are the same, the time dilation we observe in his galaxy will only equal the time dilation he observes in ours. We will both grow old at the same rate, however fast we travel, and the universe will change at the same speed for both of us.

Any experiment to decide the matter will be carried out with clocks rather than twins. It will be important that both clocks are subject to the same g forces for the duration of the experiment—remember, we are testing *Special*, not General Relativity. If both clocks are in zero g for much of the time, this will still be alright; the important thing is that both are the same except for the speed and distance they travel. Atoms that weigh the same should behave the same, unless there is something very odd going on. The clocks would have to be atomic clocks in order to measure the minute relativistic time discrepancies associated with even the fastest modern space journeys, where speeds are typically less than one ten-thousandth of the speed of light. The piffling time warp associated with accelerations of around 1 g is almost unmeasurable.

There is another paradox concerned with Special Relativity which might be called 'the mass-time paradox.' Special Relativity slows clocks. One of the earliest and simplest clocks was the rotation of a planet (the Earth), and its annual journey around a star (the sun). Like all other clocks, this one too runs more slowly as seen by a traveller in relativistic uniform motion.

But just as relativistic motion increases the mass of charged particles in the laboratory it also increases the mass of the sun, which is made of charged particles. Gravity is proportional to mass, so in order to prevent the Earth being drawn into the more massive sun the solar 'clock' should be speeded up and the Earth move faster in its orbit. It doesn't. On the contrary, a traveller passing at almost the speed of light would observe the Earth to be slowed down rather than speeded up, while we ourselves would observe no change. The apparent slowing of solar time would therefore not be matched by an apparent increase in solar gravitational mass (as opposed to solar inertial mass). Gravitational mass would appear to decrease.

The only observer to see the Earth as speeded up would be one observing from within a powerful gravitational field where his clocks run slow and all external clocks appear to run faster. So once again we have Special Relativity producing paradox while General Relativity does not. The paradox is not noticeable in laboratory experiments with sub-atomic particles, as on an atomic scale the gravitational force is unobservable and completely overwhelmed by the far more powerful electromagnetic force. In these laboratory experiments, however, *inertial* mass does appear to change.

Just as sub-atomic particles cannot combine in a nucleus without shedding a certain amount of mass, neither can they attain any fraction of the speed of light without gaining a certain amount. This applies regardless of whether they are accelerated as particles in an accelerator or as particles in the body of an astronaut. But mass cannot just be conjured out of thin air: it must be created from energy.

The energy input to an object accelerated at 1 g is very feeble in terms of mass increase, and in the course of a whole year is sufficient to generate only a tiny amount of extra mass. So this amount of acceleration, far from gaining velocity close to the speed of light as is often asserted, will in fact produce speeds which are nearer to one per cent of c! Because $E=mc^2$ the remaining energy, though a large fraction of total energy, is entirely absorbed in the production of mass which is only a tiny fraction (about one percent of one percent) of total mass. The mass increase may be trivial, but the energy required to create it is not.

Now if you could increase acceleration to many thousands of g (as is done in particle accelerators) it would be a different story, but because of the limitations of the human frame it would be dangerous to accelerate a spaceship to substantially more than 1 g for any length of time. There is no reason to believe that any other intelligent life form could endure accelerations of thousands of g (if such rocket technology were possible, which it almost certainly is not), so aliens too are doomed like us to creep around the galaxy at snail's pace. To attain anything more than the most minute fractions of c would take many years of acceleration and would require massive radiation shielding.

The conservation of mass-energy law tells us that for every ounce of mass generated an equivalent amount of energy must be injected into the system according to the equation $E=mc^2$, and for every unit of energy generated a corresponding amount of mass must disappear. No ifs or buts—it must happen and is seen to happen in particle reactions. Because for every action there is an equal but opposite reaction, only half the total energy is available to accelerate either a particle or a spaceship, and as no engine, even in theory, can ever be a hundred percent efficient, this means that for every ounce of relativistic mass increase added to the spaceship, more than two ounces of fuel mass would have to be annihilated. Not burned—*annihilated*!

It does not say much for the learned philosophers who have been asking "Where is everybody?" and surmising we are alone in the universe, that they have been unable to figure this out for themselves. And, sadly, not even relativistic time dilation and distance contraction can come to our aid, for at less than about half the speed of light distance contraction, like mass increase, is trifling. Huge time warp/length contraction effects do not arise until you are doing about three quarters of c. So now we know where everybody is: for all their advanced technologies, they are as nonplussed at the problem of bridging the void as we are!

N.B. 1 g = Earth gravity (32 feet per second per second). By a *Newtonian* calculation this rate of acceleration would indeed reach almost the speed of light in a year, but fails by millions of megatons to produce the relativistic mass-energy that would enable a spaceship to reach this velocity within so short a time (see mass increase diagram). The time dilation produced by an acceleration of 1 g has been measured with atomic clocks, and amounts to only about one fiftieth of a second per year.

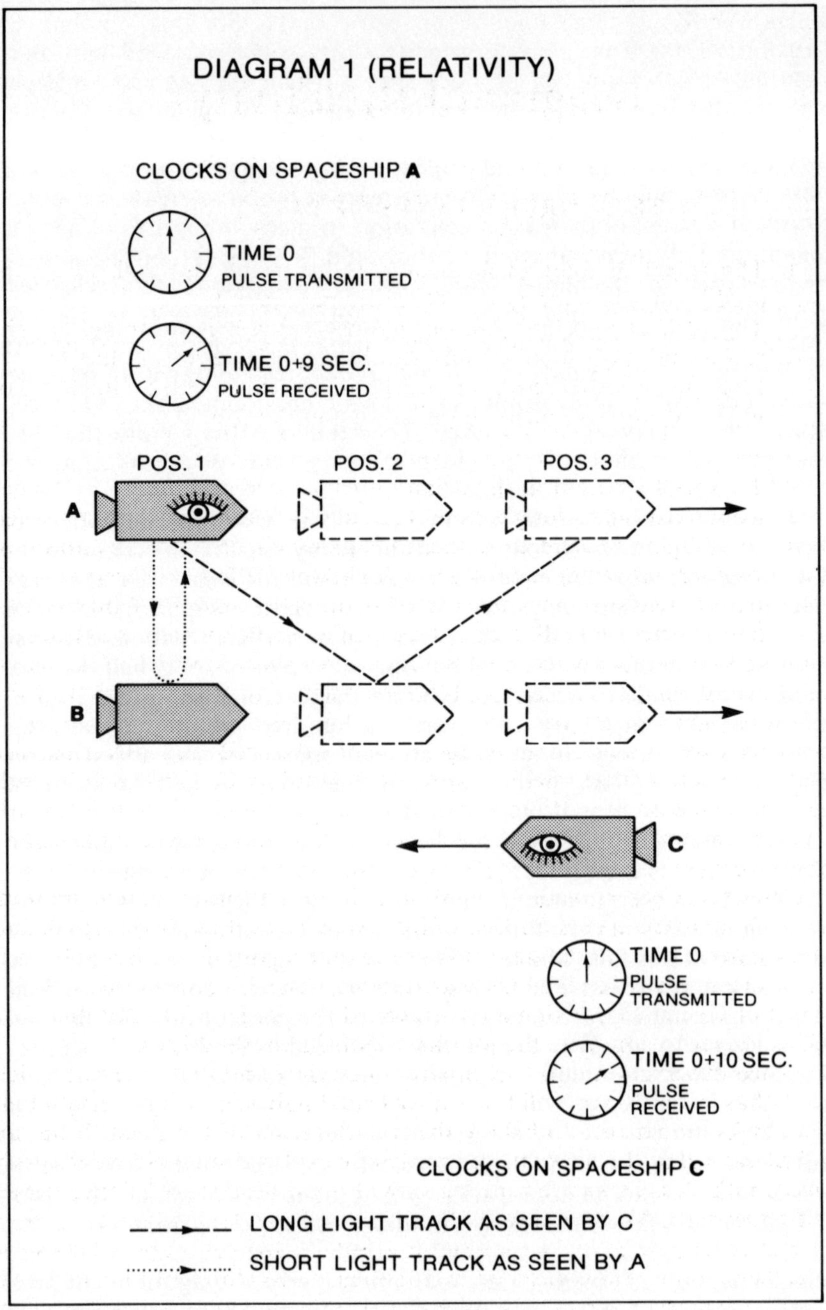
DIAGRAM 1 (RELATIVITY)
CLOCKS ON SPACESHIP A
TIME 0
PULSE TRANSMITTED
TIME 0+9 SEC.
PULSE RECEIVED
POS. 1
POS. 2
POS. 3
A
B
C
TIME 0
PULSE TRANSMITTED
TIME 0+10 SEC.
PULSE RECEIVED
CLOCKS ON SPACESHIP C
LONG LIGHT TRACK AS SEEN BY C
SHORT LIGHT TRACK AS SEEN BY A

Diagram 1 (Relativity)

THE THREE PILLARS OF RELATIVITY

1. The speed of light in a vacuum is invariant and will always be measured by an observer as 186,000 miles per second.
2. The idea of absolute motion or absolute rest is meaningless. The only rest or motion which has any validity is relative rest or motion.
3. Phenomena observed in an accelerating vehicle will be identical or near-identical to phenomena observed in a gravitational field.

If the speed of light is invariant then in order to travel a longer distance it must take a longer time. The snag is that the distance travelled by a ray of light (and therefore the time taken) depends on the motion of the observer: in other words it is relative. Different observers will measure different distances and therefore different times for light to make the same journey A to B. For a frame of reference which is stationary relative to the light source, time will appear to slow down when measured by an observer passing at speeds which are a significant fraction of the speed of light.

Spaceships A and B, moving at identical speed and direction, are travelling left to right with respect to spaceship C. Observer A sends a laser signal to B at time 0 hrs. Because he sees B as stationary, his light beam travels the short track and returns (after reflection) in time 0+9 sec.

Observer C is travelling right to left in uniform relative motion, moving at nearly half the speed of light with respect to spaceships A and B. He sees A transmit a light pulse at position 1 which is reflected from B at position 2 and received back at A at position 3. So from observer C's point of view the laser signal has travelled the longer route and therefore takes longer to complete the journey, returning in 0+10 sec.

When observers A and C compare times, they find that an event which for C has lasted 10 sec. will for A have lasted only 9. C will be certain that his clocks are correct and show the true duration of the event. If he can persuade A that this is so, there is only one explanation which will satisfy them both: A's clocks are running slow. For while C's clocks have ticked off 10 seconds, A's have ticked off only 9.

Not only are A's clocks running slow but all changes on board his ship, whether atomic, biological or mechanical, have slowed to an identical degree (as measured by C). In other words, A's time has dilated. The effect is reciprocal, and a similar experiment where the laser is fired from spacecraft C would yield the opposite result (as viewed from A or B).

DIAGRAM 2 (RELATIVITY)

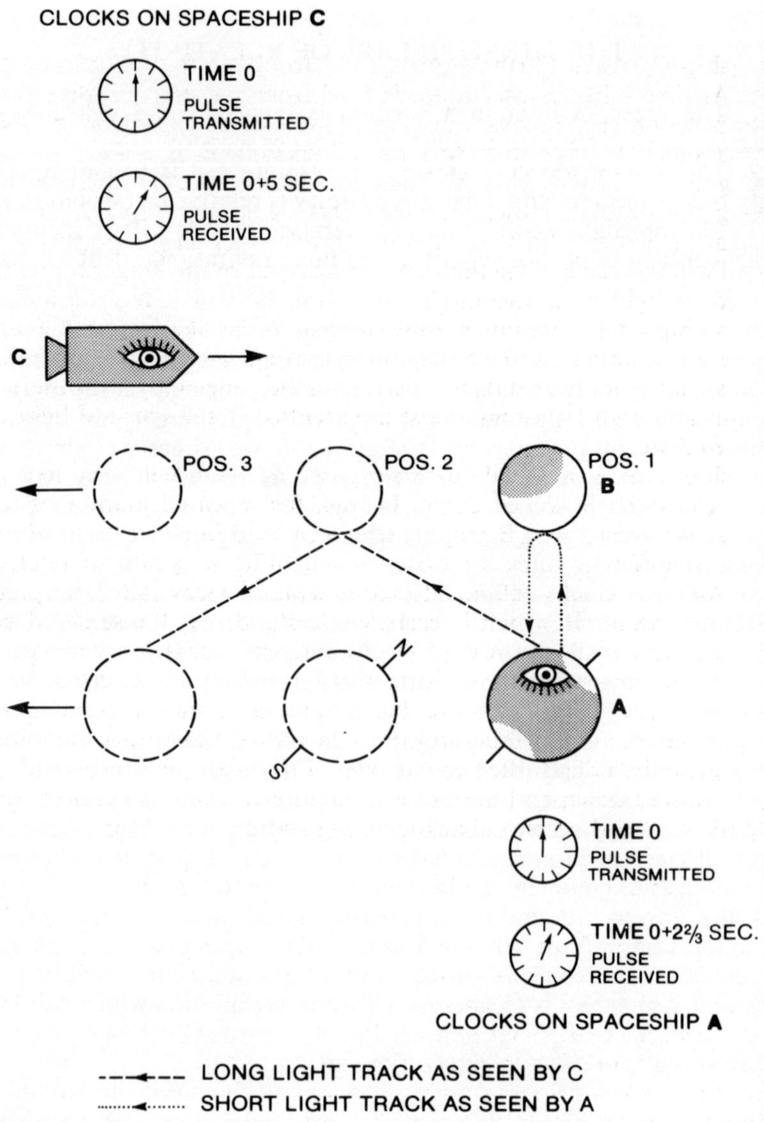

Diagram 2 (Relativity)

Spaceship C passes Earth-Moon system from left to right (South to North). At time 0 hrs. a laser beam is fired from spaceship Earth (A) to spaceship Moon (B). Because observer A sees B as stationary he will judge his signal to take the short route, reaching B at $0+1\frac{1}{3}$ sec. and returning at $0+2\frac{2}{3}$ sec. (this has actually been done, so there is no disputing the figures). However, observer C sees spaceships A and B passing at velocities which are a considerable fraction of the speed of light and will therefore judge the signal to have taken the longer route. As the speed of light in a vacuum is invariant, he will judge the longer journey to have taken a longer time. By his clocks the laser pulse will complete its journey in $0+5$ sec. So an event which we (as Earth observers) judge to have taken $2\frac{2}{3}$ sec. will be judged by an observer passing us at a high fraction (almost nine tenths) of the speed of light to have taken 5 sec.

As each of our seconds will last nearly two of his, it will seem to him that our clocks have slowed down. Indeed, the repeated journey of the light pulse between A and B may be seen as a kind of clock itself, which will beat at different rates for observers in different states of relative motion. All other clocks will be affected in a similar way. The laser pulse makes a neat example which is easily understood, but the same effects would be observed in any sort of electromagnetic radiation emanating from an object in relativistic uniform motion. For example, if the event timed was not the journey of a laser beam but a radio transmission of our pulse rates, astronaut C would find the transmitted heartbeats sounding, on average, only half as often as his own. Television pictures would be similarly slowed down and the voices of actors and presenters deep and slurred like a recording played at the wrong speed.

Glossary (The Atom)

ATOM—the smallest unit into which an element can be divided without losing its identity. Further division would turn it into atoms of lighter elements or a collection of sub-atomic particles. Consists of a massive nucleus containing one or more positive charges in the form of protons and similar but electrically neutral particles called neutrons. The nucleus is surrounded by a cloud of orbiting electrons, one for each proton in the nucleus. Electrons are of negligible mass but carry the same amount of charge as a proton, though of opposite (negative) sign.

NUCLEUS—massive central core of atom made up of protons and neutrons. It is therefore positively charged. Though containing nearly all the mass, the nucleus occupies only a negligible fraction of the atom's total volume.

PROTON—stable sub-atomic particle with mass 1836 times mass of electron. Carries positive charge equal but opposite to that of negatively charged electron.

NEUTRON—electrically neutral sub-atomic particle 1839 times mass of electron. Unstable outside nucleus. Decays into proton plus electron plus anti-neutrino, with half life of $12\frac{1}{2}$ minutes.

ELECTRON—stable and negatively charged particle which circles the positively charged nucleus in planetary orbit, held by electric forces rather than gravity. May be created by materialisation of gamma-ray photons, by particle collisions, by radio-active decay (beta particle), and the decay of muons, pions and tau particles. The much heavier muon and tau particles have sufficient electron-like properties that it has been suggested they are in some way only excited states of the electron. What is more certain is that these very short-lived particles decay by emission of neutrinos to finally become electrons.

POSITRON—positively charged electron (anti-electron) produced in materialisation of gamma rays, in a rare type of beta decay, in particle collisions, nuclear fusion and the decay of positively charged muons, pions and tau particles. On meeting an electron, both particles annihilate to become two gamma-ray photons.

PHOTON—the particle of light and other electromagnetic radiations. Only those of gamma wavelength have enough energy to materialise into electron anti-electron pairs. Has no anti-particle and is emitted by matter and anti-matter alike.

NEUTRINO—massless particle of energy which moves at the speed of light. Seldom interacts with other particles and therefore constitutes the most penetrating radiation known to science, but is harmless to living tissue. There are at least three types of neutrino plus their anti-particles: the electrons's neutrino (produced in decays which also produce electrons), the muon's neutrino (produced in the decay of muons and pions), and the tau neutrino (produced in the decay of tau particles).

RADIOACTIVE DECAY—process by which unstable atoms achieve stable form, converting to different elements on the way. Usually a combination of beta decay in which a neutron emits an electron to become a proton, and alpha decay in which an alpha particle (helium nucleus) containing two protons and two neutrons is emitted. Half the atoms in a given sample decay in a certain characteristic time (different for each element or isotope) which is known as the half life.

ALPHA RADIATION—emission of alpha particles (helium nuclei) from larger nuclei in radioactive decay. Not very penetrating, but large size and double charge of particles makes them very damaging to living tissue.

BETA RADIATION—emission of beta particles (electrons) from the nucleus in radioactive decay. Not very penetrating. Neutron emits beta particle plus anti-neutrino to become a proton, thus converting the atom to an element one place higher in the periodic table.

GAMMA RADIATION—emission of high energy electromagnetic radiation from the nucleus during alpha or beta decay and during fission or fusion. Very penetrating and very damaging to living tissue.

$E = MC^2$—energy (in ergs) = mass (in grammes) × speed of light squared, where C = speed of light in centimetres per sec. Famous equation worked out by Einstein in his 1905 Special Theory of Relativity which showed matter to be nothing more than a highly condensed form of energy (in the largest multi-megaton hydrogen bomb for example, only a few ounces of matter are converted to energy!).

ANTI-PARTICLE—most particles have 'mirror image' counterparts identical in all respects except for a few properties such as spin or charge, which are equal but opposite. The anti-particles of those forming matter can be assembled to form anti-matter. When particle meets anti-particle, both are annihilated to become energy (mainly electromagnetic).

ION—atom which has lost or gained an electron or electrons, giving a net positive or negative charge.

MOLECULE—two or more atoms linked together by electrostatic forces (essentially the magnetic attraction of opposite charges). May be composed of one element only or a number of elements (compound). The largest molecule is millions of times larger than the smallest atom.

ELEMENT—substance having a characteristic and unvarying number of protons in the nucleus of its atoms and therefore similar numbers of electrons, which gives it unique and characteristic chemical properties.

ISOTOPE—variety of an element having same number of protons in the nucleus as other varieties of the element, but a different number of neutrons. As all the isotopes of a given element have the same number of electrons, their chemical properties are the same. Most elements are found in nature as a combination of two or more isotopes.

NUCLEAR FISSION—splitting of neutron-rich heavy but unstable atoms, usually by bombardment with free neutrons, but spontaneous fission also occurs. In some elements fission can be induced by bombardment with other particles such as deuterons (deuterium nuclei), but not in a way which produces a chain reaction. An interesting type of decay has recently been discovered which is intermediate between alpha decay and fission. In radium 223 (a stage in the decay sequence of U235) a carbon nucleus is sometimes emitted rather than an alpha particle.

CHAIN REACTION—process in which neutrons emitted from an atomic nucleus cause more than one fission in adjacent nuclei, which by emitting more neutrons cause the number of fissions to build up rapidly.

CRITICAL MASS—mass of fissionable material sufficiently large (by virtue of volume to surface area ratio) to prevent the escape of too many free neutrons and so allow a self-sustaining nuclear chain reaction to take place.

THERMONUCLEAR FUSION—building up of heavier nuclei by collision of smaller ones such as protons (hydrogen nuclei). In order to overcome the mutual repulsion of their positive charges the nuclei to be fused must collide at the high speeds resulting from very high temperatures. The resulting heavy nucleus requires less mass than the sum of its constituents, and this surplus mass is emitted as electromagnetic and kinetic energy (gamma rays, neutrinos and positrons).

MEGATON—unit of energy equivalent to the detonation of a million (1,000,000) tons of T.N.T.

ERG—extremely minute unit of energy (equivalent to a snowflake striking the ground) which, however, when multiplied by the speed of light squared in centimetres per second becomes the equivalent of a small atomic bomb.

QUANTUM THEORY—highly successful theory which deals with matter and energy on a sub-atomic scale and depends on the fact that although individual particles do not behave exactly as we would expect them to by analogy with macroscopic objects, in large numbers their behaviour can be predicted statistically. While there is little doubt that the world of the sub-atomic is weird, some interpretations of the theory make it much more weird than others. The more surrealist versions have the greatest appeal to human nature—a circumstance which renders them suspect. Claims that macroscopic objects can occasionally exhibit strange quantum-like behaviour which flouts the laws of classical physics are mistaken, and there is not the slightest evidence to support such views.

Glossary (Astronomy)

QUASAR—(quasi-stellar radio source). Very young galaxy (usually very distant) in which a mysterious and super-energetic galactic centre outshines the rest of the galaxy, giving the effect of a point (quasi-stellar) source at great distances. Nearby and evolved galaxies, including our own, have these energetic centres in a more subdued form, and there has been speculation that these compact energy sources are super-massive black holes.

GALAXY—ovoid, lens or discus shaped system of stars, often forming a spiral pattern. Typically 100,000 light years across, a galaxy contains several billion stars and enormous clouds of gas and dust in rotation about a common centre of gravity. The Milky Way is our own galaxy (a spiral) seen edge on. So-called irregular galaxies are usually detached wisps of spirals.

LIGHT YEAR—distance travelled by light in one year, about six million million miles. The nearest star (other than the sun) is nearly four and a half light years away.

BLACK HOLE—quantity of matter of sufficient mass and density to produce a gravitational field where the escape velocity exceeds 186,000 miles per second (the speed of light).

WHITE HOLE—hypothetical destiny of matter drawn into a black hole singularity. It is speculated that the warping of time and space may be sufficient to eject the vanished matter as a flood of energy in some other part of the universe.

SINGULARITY—black hole produced by matter collapsed to a state of infinite density (denser than a neutron star). It is thought that stars of more than about 10 solar masses will not expel sufficient mass to end their days as neutron stars, and when no longer supported by radiation pressure the remnant cores of the resulting supernovae will be crushed to a point source where matter is in a state not fully understood by science. Stellar remnants of less than 3 solar masses are too small to form singularities and will become neutron stars or white dwarves.

NEUTRON STAR—remnant of massive star which has undergone a supernova explosion, throwing off the outer layers and crushing the core into a super-dense state in which electrons are fused with protons to form neutrons. This is the most extreme density obtainable by matter as we know it, and resembles a gigantic atomic nucleus perhaps only 6 miles in diameter but containing material equivalent to more than the entire mass of the sun (up to 3·2 solar masses).

PULSAR—rapidly rotating neutron star emitting twin beams of electromagnetic radiation which pulse or flash like a lighthouse each time one of them is aligned in the direction of Earth.

WHITE DWARF—highly condensed stellar remnant about the size of Earth but similar in mass to the sun. This is the ultimate fate of small sun-like stars which have insufficient mass to end as supernovae. Up to about $1\frac{1}{2}$ solar masses the pressure of a 'degenerate' electron gas is sufficient to halt further collapse, but larger stellar remnants will overcome this 'electron degeneracy pressure' and shrink dramatically to become neutron stars or black holes. White dwarves are very hot (hence their colour) but owing to their small size give off very little radiation.

PLASMA—electrically neutral, very hot, and sometimes very dense gas composed of charged particles rather than atoms or molecules. The material of stars is mainly in the form of plasmas, which at the centres of stars can be very much denser than lead.

Glossary (Biology)

NUCLEOTIDE—strictly speaking, a combination of one of the four bases (adenine, thymine, guanine and cytosine, or their variants) with a ribose and phosphate side chain group, but in more casual use denotes the base itself.

NUCLEIC ACID—chain of nucleotides linked at the sides by alternate ribose and phosphate molecules, forming DNA or RNA. The variable sequence of nucleotides along the chain dictates the structure of proteins and enzymes to be synthesised, and hence the structure of the eventual living organism.

BASE PAIR—pair of bases or nucleotides linked in the middle by hydrogen (electrostatic) bonds, forming a rung or step in the nucleic acid chain and connected to other pairs by ribophosphate side chains. The adenine-thymine combination forms one pair and guanine-cytosine the other. Because of differences in electrostatic bonding, size and shape, members of one base pair cannot exchange partners with members of the other. Adenine must always pair with thymine (or vice versa), and guanine must always pair with cytosine.

AMINO ACID—molecules composed mainly of carbon and nitrogen which are the building blocks of proteins and enzymes. There are twenty main types of amino acid found in biological structures (plus two more which are not used in proteins and have only recently been discovered).

PROTEIN—chains of amino acids which form the main structural materials of living organisms.

ENZYME—type of protein whose main function is catalytic (promotes chemical reactions) rather than structural. Enzymes are large, flexible molecules which adapt to the shape of certain other molecules and alter their electrostatic bonding, thus promoting a chemical reaction.

ELECTROSTATIC BONDING—means by which atoms are joined to make molecules and small molecules joined to make larger ones. Essentially the magnetic attraction of opposite charges. Unlike human artifacts, which are joined together with nails, screws, rivets etc., the microscopic artifacts of biology must be joined, in effect, by electromagnets.

DE-OXY-RIBONUCLEIC ACID (DNA)—chemical chain or ladder of helical configuration with variable sequence of nucleotide sub-units in which are encoded the genetic instructions for replication of the parent organism. As the name suggests, the ribose (sugar) sub-units which form the side chains are lacking an oxygen atom, and therefore differ from those found in RNA.

RIBONUCLEIC ACID (RNA)—nucleic acid very similar to DNA, but containing ribose sub-units of slightly different structure, and the variant uracil replaces thymine in the partnership with adenine. In most organisms RNA is relegated to a subsidiary 'messenger' role, transcribing instructions from the DNA genetic 'master tape', but in some viruses it replaces DNA completely.

ISOMER—many molecules (including some of biological significance) exist in two identical but mirror image forms showing 'left handed' and 'right handed' symmetry. Amino acids of non-biological origin contain equal amounts of both forms, but in biological structures only the 'left handed' isomer is used.

CODON—three letter 'word' in the genetic 'message' which specifies a particular amino acid for the build-up of polypeptide chains (proteins and enzymes). The 'letters' are the four nucleotides, adenine, thymine, guanine and cytosine, which form the 'alphabet'. CCG for example, codes for the amino acid proline.

GENE—a sentence in the genetic message giving instructions for some physical characteristic in living organisms. The 'sentence' may be long or short and therefore varies in complexity and the number of codons required.

CHROMOSOME—long segment of DNA containing many genes, coiled and supercoiled around protein sub-units in order to get the maximum possible information into the smallest possible space. Man has 46 chromosomes arranged in 23 pairs—fewer than some lower animals or even plants, so the number of chromosomes is not always proportional to the number of genes or the complexity of the genetic message. Chromosomes bearing the complete instructions for replicating the entire parent organism are found in *every* cell nucleus. The reproductive cells have only half the normal requirement so that when they fuse at the moment of conception the full number will be present, with one of each chromosome pair donated by either parent.

PROCARYOTE—primitive bacteria-like unicellular organism having no well defined nucleus. The DNA genetic instruction 'tapes' form a loose clump in the intracellular fluid, and the specialised organelles (metabolic structures) of higher organisms are virtually absent.

EUCARYOTE—larger and more complex organism similar to the cells of higher animals (which are also eucaryotes), having well defined nucleus enclosed in a membrane, and several large organelles.

METABOLISM—chemical processes necessary for life in which simple molecules are built up into larger ones and vice versa.

RECOMBINATION (CROSSING OVER)—during cell division sections of DNA, assisted by enzymes, can occasionally change places from one member of a chromosome pair to another, so that genetic characters that were formerly linked become uncoupled and new linkages established—in effect a reshuffling of the genetic cards and perhaps even, on rare occasions, the creation of a completely new card for introduction into the pack.